ASCE Manuals and Reports on Engineering Practice No. 133

AF718386

# Pilot Tube and Other Guided Boring Methods

Prepared by
Task Committee on Pilot Tube and Other Guided Boring Methods of the
Committee on Trenchless Installations of Pipelines of the
Utility Engineering and Surveying Institute of the
American Society of Civil Engineers

ASCE AMERICAN SOCIETY OF CIVIL ENGINEERS

UESI UTILITY ENGINEERING & SURVEYING INSTITUTE

Published by the American Society of Civil Engineers

**Library of Congress Cataloging-in-Publication Data**

Names: Utility Engineering and Surveying Institute (American Society of Civil Engineers). Task Committee on Pilot Tube and Other Guided Boring Methods, author.
Title: Pilot tube and other guided boring methods / prepared by Task Committee on Pilot Tube and Other Guided Boring Methods of the Committee on Trenchless Installations of Pipelines of the Utility Engineering and Surveying Institute of the American Society of Civil Engineers.
Description: Reston, Virginia : American Society of Civil Engineers, [2017] |
Series: ASCE manuals and reports on engineering practice ; no. 133 |
Includes bibliographical references and index.
Identifiers: LCCN 2017010889| ISBN 9780784414743 (paperback : alk. paper) |
ISBN 9780784480571 (PDF) | ISBN 9780784480663 (ePUB)|
Subjects: LCSH: Underground pipelines–Equipment and supplies. | Tubes. | Underground pipelines–Design and construction. | Directional drilling–Equipment and supplies. |
Trenchless construction. | Boring.
Classification: LCC TJ933 .P48 2017 | DDC 621.8/672–dc23 LC record available at https://lccn.loc.gov/2017010889

Published by American Society of Civil Engineers
1801 Alexander Bell Drive
Reston, Virginia, 20191-4382
www.asce.org/bookstore | ascelibrary.org

Any statements expressed in these materials are those of the individual authors and do not necessarily represent the views of ASCE, which takes no responsibility for any statement made herein. No reference made in this publication to any specific method, product, process, or service constitutes or implies an endorsement, recommendation, or warranty thereof by ASCE. The materials are for general information only and do not represent a standard of ASCE, nor are they intended as a reference in purchase specifications, contracts, regulations, statutes, or any other legal document. ASCE makes no representation or warranty of any kind, whether express or implied, concerning the accuracy, completeness, suitability, or utility of any information, apparatus, product, or process discussed in this publication, and assumes no liability therefor. The information contained in these materials should not be used without first securing competent advice with respect to its suitability for any general or specific application. Anyone utilizing such information assumes all liability arising from such use, including but not limited to infringement of any patent or patents.

ASCE and American Society of Civil Engineers—Registered in U.S. Patent and Trademark Office.

*Photocopies and permissions.* Permission to photocopy or reproduce material from ASCE publications can be requested by sending an e-mail to permissions@asce.org or by locating a title in ASCE's Civil Engineering Database (http://cedb.asce.org) or ASCE Library (http://ascelibrary.org) and using the "Permissions" link.

*Errata:* Errata, if any, can be found at http://dx.doi.org/10.1061/9780784414743

Copyright © 2017 by the American Society of Civil Engineers.
All Rights Reserved.
ISBN 978-0-7844-1474-3 (paper)
ISBN 978-0-7844-8057-1 (PDF)
ISBN 978-0-7844-8066-3 (ePUB)
Manufactured in the United States of America.

21 20 19 18 17 1 2 3 4 5

# MANUALS AND REPORTS ON ENGINEERING PRACTICE

(As developed by the ASCE Technical Procedures Committee, July 1930, and revised March 1935, February 1962, and April 1982)

A manual or report in this series consists of an orderly presentation of facts on a particular subject, supplemented by an analysis of limitations and applications of these facts. It contains information useful to the average engineer in his or her everyday work, rather than findings that may be useful only occasionally or rarely. It is not in any sense a "standard," however; nor is it so elementary or so conclusive as to provide a "rule of thumb" for nonengineers.

Furthermore, material in this series, in distinction from a paper (which expresses only one person's observations or opinions), is the work of a committee or group selected to assemble and express information on a specific topic. As often as practicable the committee is under the direction of one or more of the Technical Divisions and Councils, and the product evolved has been subjected to review by the Executive Committee of the Division or Council. As a step in the process of this review, proposed manuscripts are often brought before the members of the Technical Divisions and Councils for comment, which may serve as the basis for improvement. When published, each work shows the names of the committees by which it was compiled and indicates clearly the several processes through which it has passed in review, so that its merit may be definitely understood.

In February 1962 (and revised in April 1982), the Board of Direction voted to establish a series titled "Manuals and Reports on Engineering Practice," to include the Manuals published and authorized to date, future Manuals of Professional Practice, and Reports on Engineering Practice. All such Manual or Report material of the Society would have been refereed in a manner approved by the Board Committee on Publications and would be bound, with applicable discussion, in books similar to past Manuals. Numbering would be consecutive and would be a continuation of present Manual numbers. In some cases of joint committee reports, bypassing of Journal publications may be authorized.

*A list of available Manuals of Practice can be found at http://www.asce.org/bookstore.*

# CONTENTS

# PREFACE

This manual of practice was prepared by the Task Committee on Pilot Tube and Other Guided Boring Methods of the ASCE Committee on Trenchless Installation of Pipelines (TIPS), as part of the Utility Engineering & Surveying Institute (UESI). The manual describes the current pilot tube and other guided boring methods used by engineers and construction professionals in designing and installing pipelines to a design line and grade under roads, railroads, streets, and other constructed and natural structures and obstacles.

This manual of practice has been created by a group of engineers, owners, suppliers, manufacturers, and contractors fully knowledgeable of the method and its use. This manual takes into account many of the advances that have occurred over the years in the guiding of trenchless methods. Many of the sections provide a summary of the state of the industry as of 2016. The task committee acknowledges that the technology continues to change and that changes in construction continue to develop.

Sections have been written assuming the reader may be new to the various construction methods included in this manual. No document including this one can encompass all of the issues on a particular pilot tube or other guided boring project. Improvements in best practices and technology continue to evolve so quickly that consideration of this manual on any project must take into account not only the specific characteristics of the particular project but also further improvements in best practices and technology.

The engineer of a pipeline project is encouraged to consider all trenchless methods before concluding that the pilot tube and/or other guided boring methods are the most suitable construction methods available. Manuals and reports on engineering practice (known as *MOPs*) have been written by ASCE for different construction methods. A list of useful references is provided at the end of Chapter 1.

If the engineer responsible for the pipeline project does not have a strong background in trenchless design, an engineering firm that specializes in trenchless designs should be consulted to provide a peer review early in the planning/design process to help ensure good design choices are being made.

# ACKNOWLEDGMENTS

The American Society of Civil Engineers (ASCE) and the Utility Engineering & Surveying Institute (UESI) acknowledge the work of the Task Committee on Pilot Tube and Other Guided Boring Methods. This group comprises individuals from many backgrounds, including design and consulting engineering, the construction industry, equipment and pipe manufacturing, academia, and government.

Principal authors:
Jeffrey J. Boschert, P.E., *Task Committee Secretary*
Glenn M. Boyce, Ph.D., P.E.
Dennis J. Doherty, P.E.
Andrew Finney, P.E., G.E., P.Eng.
Mohammad Najafi, Ph.D., P.E., *Task Committee Chair*
Richard Palmer, P.E.
Troy Stokes
Nick Strater, P.G.

Other individuals who served and contributed:
Don Bergman
Richard Botteicher
Mark Bruce
Ralph Carpenter
David Crandall
Brian Dorwart, P.E.
Jason Holden, P.E.
Daniel L. Liotti, P.E.
Shah Rahman
Rick Turkopp, P.E.
Camille Rubeiz

Thank you to the Blue Ribbon Review Committee who reviewed the finished document and provided their comments:

Jason Lueke, Ph.D., P.Eng., *Chair*
Maynard Akkerman
Alan Atalah, Ph.D., P.E.
Lester M. Bradshaw Jr.
Craig Camp

Thank you to Julie McCullough for providing technical editing, Niloofar Rezaei and Nader Mavaddat for editorial review, and Laura Anderson, Yovani Zelaya, and Jon Clugston for graphic support.

The task committee would like to acknowledge the support provided by the following companies, association, and university:

Akkerman
Brierley Associates
CUIRE at the University of Texas at Arlington
CH2M
Haley & Aldrich
McMillen Jacobs Associates
National Clay Pipe Institute
Northeast Remsco Construction, Inc.

ASCE Committee on Trenchless Installation of Pipelines (TIPS):

Chair: Jeffrey J. Boschert, P.E.
ExCom Liaison: Tennyson Muindi, P.E.

# ABBREVIATIONS AND ACRONYMS

| | |
|---|---|
| AAA | American Arbitration Association |
| ADR | alternative dispute resolution |
| AIME | American Institute of Mining, Metallurgical and Petroleum Engineers |
| ANSI | American National Standards Institute |
| API | American Petroleum Institute |
| ASCE | American Society of Civil Engineers |
| ASTM | American Society for Testing and Materials |
| AWWA | American Water Works Association |
| BVR | boulder volume ratio |
| CCTV | closed-circuit television |
| CFR | Code of Federal Regulations |
| CPT | cone penetration testing |
| CVR | cobble volume ratio |
| DI | ductile iron |
| DM | design memoranda |
| DMP | deformation monitoring points |
| DRB | dispute review boards |
| DIN | Deutsches Institut für Normung; in English, the German Institute for Standardization |
| DSC | differing site conditions |
| EBR | environmental baseline report |
| EN | European Norme; in English, European Standard |
| EPDM | ethylene propylene diene monomer |
| FEMA | Federal Emergency Management Agency |
| FHWA | Federal Highway Administration |
| FRPM | fiberglass-reinforced polymer mortar |
| GBR | geotechnical baseline report |

| | |
|---|---|
| GDM | geotechnical design memorandum |
| GDR | geotechnical data report |
| GDSR | geotechnical design summary report |
| GIR | geotechnical interpretative report |
| GSI | geological strength index |
| HDD | horizontal directional drilling |
| HDPE | high-density polyethylene |
| ID | inside diameter |
| IJS | intermediate jacking stations |
| ISRM | International Society for Rock Mechanics |
| LED | light-emitting diode |
| MOP | Manual and Report on Engineering Practice |
| MSDS | material safety data sheets |
| MUTCD | Manual of Uniform Traffic Control Devices |
| NASTT | North American Society for Trenchless Technology |
| NSF | National Sanitation Foundation |
| OD | outside diameter |
| OSHA | Occupational Safety and Health Administration |
| PC | polymer concrete |
| PCH | powered cutterhead |
| PPE | personal protective equipment |
| PRH | powered reamer head |
| PVC | polyvinyl chloride |
| Q | flow capacity or system for classification of rock mass |
| QA | quality assurance |
| QC | quality control |
| QLA | Quality Level A |
| QLB | Quality Level B |
| QLC | Quality Level C |
| QLD | Quality Level D |
| RC | reinforced concrete |
| RDA | rock drill adapter |
| RMR | rock mass rating |
| RQD | rock quality designation |
| SAT | soil abrasion test |
| SDS | safety data sheets |
| SI | International System of Units |
| SMP | structural monitoring points |
| SPT | standard penetration test |
| TBM | tunnel boring machine |
| TM | technical memoranda |
| UCS | unconfined compressive strength |
| UMP | utility monitoring points |
| VC | vitrified clay |

# CHAPTER 1
# INTRODUCTION

Utilities of various types, sizes, and purposes are installed below the surface routinely in every community around the world. The pipelines are the infrastructure that provides our homes, our office buildings, our schools, and other structures with waste removal via sanitary sewers, stormwater drainage, potable water, gas, recycled water, irrigation, electricity, communication, and cable TV (as listed in order of most common current usage).

These pipelines can be installed by a variety of construction methods, with the most common being open trench excavation. The pipelines can be found anywhere, but are typically installed within the rights-of-way of our roads, streets, and highways. As new pipelines are installed below the surface and more utilities are abandoned, the density of the underground infrastructure increases. As the amount of traffic increases on the road surfaces, access to the subsurface also decreases. Installing new infrastructure and pipelines within an existing right-of-way has become more difficult. However, not all pipelines can be installed by open trench excavation. The pipeline industry has adopted and now offers increasingly sophisticated solutions to meet the challenges of installing pipelines in congested, urban environments around the world. Because these new construction solutions/methods do not rely on digging a trench, the methods are known as trenchless methods.

Some of the more widely used trenchless methods for new utility installations are (as listed alphabetically)

- Guided boring through nondisplaceable soils or rock;
- Guided boring using a pilot tube;
- Horizontal auger boring;
- Horizontal directional drilling (HDD);

- Impact moling;
- Microtunneling;
- Pipe jacking;
- Pipe ramming; and
- Utility tunneling.

Trenchless methods are often attractive to project owners and often required by regulatory agencies and property owners because they generally result in fewer impacts to the surface, existing structures, and infrastructure than do traditional open trench excavations. Aboveground effects are reduced because only shafts, pits, or portals have to be excavated rather than a continuous trench. Underground impacts are reduced because the pipe can be installed without excavating the otherwise conflicting underground utilities or surface structures. The application of certain trenchless methods enables alignment control in established corridors and rights-of-way that are becoming more congested. In general, the benefits of trenchless construction methods can be summarized as follows:

- Safer for the public;
- Lowers social and economic effects;
- Minimizes environmental effects including carbon footprint;
- Reduces traffic effects;
- Allows for deeper installations without significant increases in the cost;
- Mitigates impacts to existing surface and subsurface infrastructure;
- Addresses potentially challenging subsurface conditions such as soft or loose soil, rock, or the groundwater table;
- Allows more freedom in alignment selection within or outside of established rights-of-way; and
- Reduces project duration or work content.

Trenchless design is much more sensitive to subsurface conditions than open trench excavation. It is usually more difficult, costly, and time-consuming to adapt trenchless methods to unexpected ground conditions than it is to adapt open trench excavations; primarily because when trenchless methods are used, the material being excavated is not visible or directly accessible. When using trenchless methods, it is critical to conduct a thorough geotechnical and subsurface investigation to determine, among other things, geology, ground type and anticipated behavior, groundwater, and the presence of buried objects placed naturally or as a remnant of past construction. The anticipated conditions affect the selection of the trenchless method, the design layout, and even the contract provisions.

The basic design process for a trenchless project should include identifying and understanding project requirements, owner design requirements, third-party requirements, ground conditions, and risks and risk tolerance. This information is then used in the design phase to select the most suitable trenchless method; provide a well-thought-out design that minimizes the risks; mitigate the controllable risks; identify potential claims and address them; and prepare contract clauses that help allocate risks and manage disputes.

The decision to use a trenchless method, and the selection of a particular method from those available, should be made methodically and systematically and be based on project requirements. Trenchless methods, equipment, and guidance systems have improved and changed over time, and contractors have continued to apply innovative new techniques to construction. Therefore, it is important for design engineers to remain current with all available trenchless methods to systematically evaluate options. Regardless of the trenchless method selected, adherence to the following guidelines can reduce project cost, schedule, and risk:

- Use uniform outside diameters (ODs) to the extent possible;
- Use similar drive lengths to the extent possible;
- Keep drive lengths within practical limits;
- Minimize the number of shafts;
- Locate shafts at typical maintenance access points; and
- Evaluate alternative alignments.

As noted already, there are a number of different trenchless methods available to engineers and contractors. These trenchless methods use different means to excavate and support the ground and install the pipeline, offering different capabilities on pipe size, pipe material, or need for a casing; drive length; installation above or below the groundwater table; installation at a precise line and grade for gravity flow within the pipeline; and installation of the pipeline directly (one-pass process) or with a casing (two-pass process).

Table 1-1 summarizes the typical pipe sizes, installation lengths, ability to install pipelines above or below the groundwater table, ability to install pipelines at a precise line and grade, and ability to install the carrier pipe directly for the more common trenchless methods. The pipe sizes and drive lengths reflect increases that have occurred as the industry has matured. However, it is important to remember that the emphasis should always be placed on managing risk first, not matching or exceeding lengths accomplished in the past. Actual and perceived risks directly increase an owner's cost.

This manual and report on engineering practice (MOP) focuses specifically on the pilot tube and other guided boring methods. The chapters in this manual will describe the following:

- Chapter 2: The pilot tube and other guided boring methods in detail;
- Chapter 3: Planning;

Table 1-1. Typical Sizes, Lengths, and Capabilities of Various Trenchless Methods

| Method | Size, OD (in.) | Length (ft) | Groundwater | Grade Control |
|---|---|---|---|---|
| Pipe Ramming[a] | 8–96 | 250 | Both[b] | No[c] |
| Horizontal Auger Boring[a] | 12–72 | 300 | Above[b] | No[c] |
| Guided Boring[a] | 4–48 | 300 | Both[b] | Yes |
| Microtunneling[d] | 30–96 | 1,500 | Both | Yes |
| Shielded Pipe Jacking[d] | 48–120 | 1,500 | Above[b] | Yes |
| Horizontal Directional Drilling | 4–42 | 6,500 | Both | No |
| Utility Tunneling[d] | 48 or larger | Any length | Above[b] | Yes |

Note: These are typical parameters and do not represent absolute minimum and maximum, nor do they consider additional ground improvement methods or other risk mitigation methods. These values are subject to change as technology evolves.

[a]Length decreases as diameter increases.

[b]Below the groundwater may be possible and is limited by ground type or depth below groundwater.

[c]Grade control may be achievable through steering heads or hybrid methods, as discussed herein.

[d]Length increases as diameter increases.

- Chapter 4: Detailed site investigations;
- Chapter 5: Shaft design;
- Chapter 6: Pipe characteristics and design;
- Chapter 7: Design and contract documents;
- Chapter 8: Construction; and
- Glossary.

For information on the other trenchless methods noted in Table 1-1, the reader may refer to the following:

- Pipe Ramming: ASCE *MOP No. 115* (2008);
- Horizontal Auger Boring Projects: ASCE *MOP No. 106* (2017);
- Microtunneling: ASCE Standard 36-15 (2015);
- Shielded Pipe Jacking: Thomson (1994);
- HDD: ASCE *MOP No. 108* (2014);
- Utility Tunneling: Proctor and White (1977); and

## REFERENCES

ASCE. (2017). "Horizontal auger boring projects." *Manuals and Rep. on Engineering Practice No. 106,* 2nd Ed., Reston, VA.

ASCE. (2007). "Pipe bursting projects." *Manuals and Rep. on Engineering Practice No. 112,* Reston, VA.

ASCE. (2008). "Pipe ramming projects." *Manuals and Rep. on Engineering Practice No. 115,* Reston, VA.

ASCE. (2014). "Pipeline design for installation by horizontal directional drilling." *Manuals and Rep. on Engineering Practice No. 108,* 2nd Ed., Reston, VA.

ASCE. (2015). "Standard design and construction guidelines for microtunneling." *ASCE 36-15,* Reston, VA.

Proctor, R. V., and White, T. L. (1977). *Earth tunneling with steel supports,* Commercial Shearing, Inc., Youngstown, OH.

Thomson, J. (1994). *Pipejacking and microtunneling,* Chapman and Hall, New York.

# CHAPTER 2

# THE PILOT TUBE AND OTHER GUIDED BORING METHODS IN DETAIL

## 2.1 INTRODUCTION TO THE PILOT TUBE METHOD

The pilot tube method is defined herein as a multistage method of accurately installing a pipe to line and grade by use of a guided pilot tube followed by upsizing to install the pipe. The pilot tube method is also referred to as the guided boring method and has been previously referred to as pilot tube microtunneling. However, the term *pilot tube microtunneling* does not meet the current industry-accepted definition of microtunneling (ASCE 36-15); therefore, it should no longer be used. The pilot tube method is a hybrid technique that combines features of different trenchless methods such as horizontal auger boring, microtunneling, and horizontal directional drilling (HDD) to accurately install a pipe to line and grade. Dependent on the application and compared with microtunneling, the pilot tube method can be a cost-effective tool for the accurate installation of small-diameter pipelines.

This method incorporates the use of a pilot tube (Fig. 2-1), which is pushed through the ground and might be followed by upsizing tools, spoil removal systems, and ultimately the carrier pipe. The pilot tube is guided using a camera-mounted theodolite. The accuracy of the pilot tube installation is based on maintaining visual contact with the light source at the end of the pilot tube. The overall accuracy is dependent on a number of factors, such as

- Subsurface conditions;
- Setup, transfer, and maintenance of survey control points;
- Equipment setup;
- Drive length; and
- Air quality/humidity inside the pilot tube.

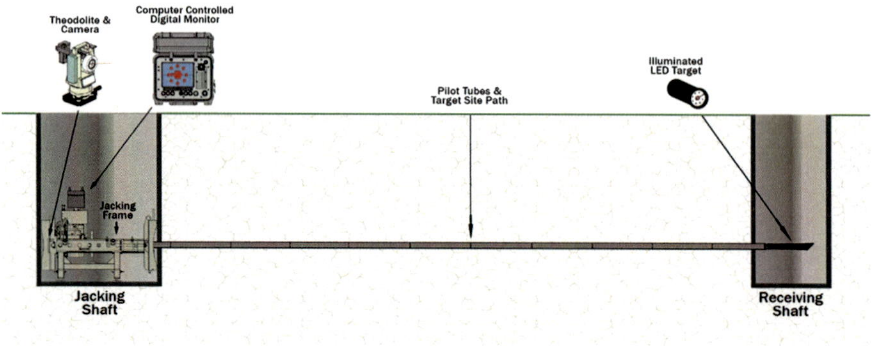


*Fig. 2-1. Installation of the pilot tube*

The pilot tube displaces the ground during advancement, thereby creating an initial small-diameter borehole. No spoils are removed during pilot tube installation. Generally, the initial borehole is then enlarged using an upsizing tool as the spoils are removed. As the enlargement of the initial borehole is made, the pilot tube is systematically advanced to and retrieved from the receiving shaft and is available for reuse on the next installation. Additional enlargement passes may be required. The final step is to jack the carrier pipe or casing into place while pushing the upsizing tools and spoil removal system into the receiving shaft for subsequent recovery. Because the pilot tube is accurately installed and all the following steps use the pilot tube or subsequent upsizing tools for guidance, the final carrier pipe can be installed with similar accuracy.

Typically, the pilot tube is installed in soils that can be compressed by displacement. Although some published documents suggest a displacement threshold of a standard penetration test (SPT or N-value) of 50, the authors believe that an N-value of 30 or less is more suitable. If the ground is too hard (N-value greater than 30) or in rock, the slanted steering head can be replaced with a drill bit that can break and remove the hard material at the leading end of the pilot tube. In these cases, the pilot tube initial borehole is made by boring into the ground, and the same guidance system is used to monitor the line and grade of the pilot tube. Gravels, cobbles and boulders can cause some challenges during construction. Recent developments, such as lubricants for loose sands, water control reaming heads for wet sands, and tooling for nondisplaceable ground, have increased the possibilities for use of the pilot tube method in ground conditions that were once considered impossible. See Chapter 4 for additional information on the geotechnical investigations needed and Table 4-2, which provides applicable ground conditions for the pilot tube method. As with any underground construction, a good geotechnical investigation is crucial to the final success of the project.

Among the most notable reasons for the popularity of the pilot tube method are its accuracy, ability to retract and reinstall, small work space requirement, and small jacking shaft requirement.

The pilot tube method was developed in Germany in the 1980s to enable the installation of 4- to 6-in. (100 to 150 mm) diameter sewer laterals (Stein, Mollers, and Bielecki, 1989). The method was first used in the United States in the late 1990s. The initial capabilities of the pilot tube method were in the range of 4- to 12-in. (100 to 305 mm) outside diameter (OD) pipes for laterals and small-diameter sewer main lines, with single drive lengths up to 250 ft (76 m). The technology has now grown to include typical installation of pipes up to 48 in. (1.2 m) in OD with drive lengths in the 300-ft (91-m) range. Improvements in optical guidance systems and greater hydraulic power in the jacking frames have made these larger diameters and longer drive lengths possible.

Since the initial uses, the method has evolved to combine elements of other techniques to enable installation of larger-diameter pipes. See Table 1.1 for typical sizes and lengths. The pilot tube is also used for

- Guiding horizontal auger bore installations;
- Guiding pipe ram installations;
- Guiding precise short HDD installations;
- Probing for buried objects that may be in the path of other trenchless installations;
- Installing small-diameter pipes horizontally for ground stabilization;
- Pulling back utilities; or
- Any application in which precise line and grade are needed in the ground.

## 2.2 INSERTING THE PILOT TUBE

Pilot tubes are hollow, stiff-walled, steel tubes that are fastened to each other via threaded or hex connections. The tubes are available in double or single wall depending on the manufacturer. On double-walled systems, a lubricant may be pumped through the annular cavity between the tubes to the steering head to reduce ground friction. On some double-walled tube systems, the inner tube will rotate with the steering head during advancement for torque reduction. These pilot tubes typically have an OD of 4 to 5 in. (100 to 125 mm) with a 2- to 3-in. (50 to 75 mm) inside diameter hollow sight path. The typical individual pilot tubes range in length from 30 to 60 in. (762 to 1,525 mm), depending on the size of the jacking frame and shaft diameter (Fig. 2-2).

At the leading end, the pilot tube is equipped with a steering head that enables the control of line and grade and is generally configured with a 45° slanted plane, a 30° slanted plane, or a bullet head (Fig. 2-3). The choice of

(a) (b)

*Fig. 2-2. Different rack configurations of pilot tubes (photos courtesy of Akkerman and Bohrtec)*

steering head depends on the anticipated ground conditions. Generally, the harder the ground (higher N-value), the lower the angle of the slanted plane.

The steering head also houses the light-emitting diode (LED) target [Fig. 2-4(a)] that is viewed with the camera-mounted theodolite [Fig. 2-4(b)]. The target has a center LED with two concentric rings of LEDs around it, and an indicator to show the orientation of the slanted face of the steering head. The LED lights are low-energy-demand lights and are battery powered. The illuminated target remains *on* during the installation of the pilot tube.

The guidance system incorporates a camera-mounted theodolite that is independently anchored in the jacking shaft and positioned to look down the center of the hollow pilot tube at the LED-illuminated target. The camera image from the theodolite is transmitted to a digital display (Fig. 2-5), giving the operator continuous information about the location and orientation of the steering head at the end of the pilot tube.

The machine operator uses the real-time information on the digital display to accurately guide the pilot tube installation. A jacking frame, set to the design line and grade, is used to thrust the pilot tubes into the ground (Fig. 2-6). The first pilot tube is pushed into the ground, displacing the ground to form a borehole. Based on the orientation of the steering head, the illuminated center LED and the two rings of LEDs will begin to deviate from the crosshairs on the digital display. After the first pilot tube is inserted, the thrust cylinders and machine are retracted, and a second pilot tube is added. To correct any deviation from line and grade, the operator rotates the steering head using the illuminated indicator for guidance. The objective is to keep the center LED in the center of the crosshairs.

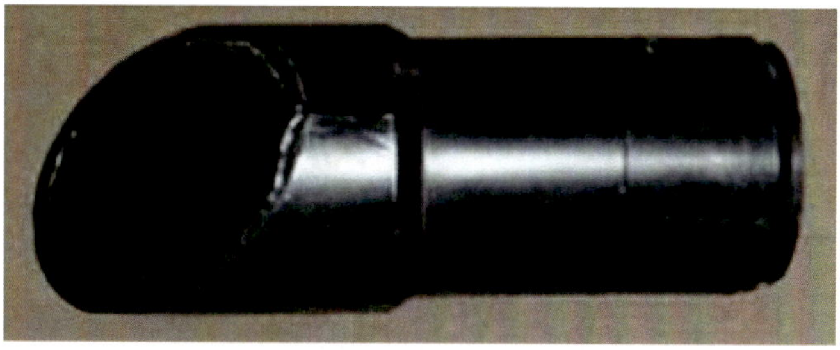

(a) 45-degree slanted head

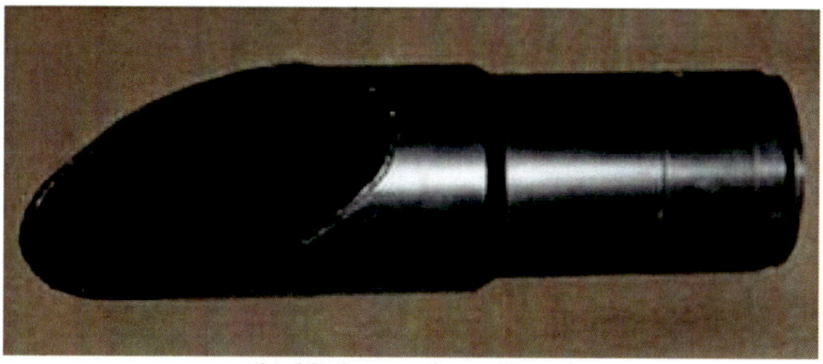

(b) 30-degree slanted head

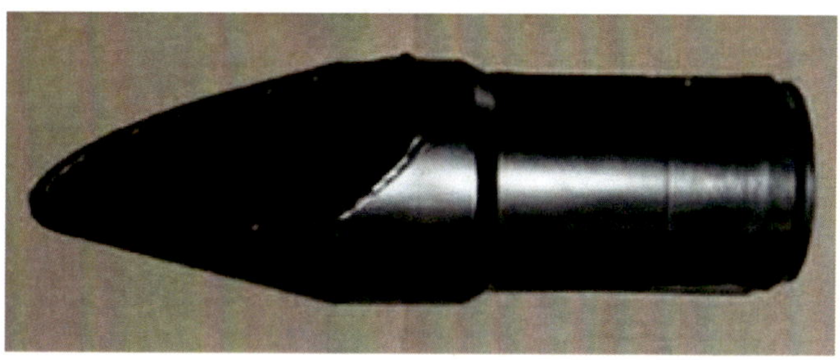

(c) Bullet head

*Fig. 2-3. Three types of slanted heads used with the pilot tube (photos courtesy of Akkerman)*

This process is repeated throughout the installation of the pilot tubes to maintain line and grade.

A typical jacking frame ranges from 6.6 to 14 ft (2 to 4.3 m) and has 100 to 265 tons of capacity. Fig. 2-7 shows an 8-ft (2.4 m) long, 100-ton frame. It can be adjusted to fit within the shaft at the required height as well as line and grade. Some frames can be extended to accommodate longer lengths of jacking pipe. The jacking frame incorporates hydraulic jacks to provide thrust for the pipe installation, transfers the jacking force to the shaft wall, and uses a motor to provide torque to the pilot tube and augers. The camera-mounted theodolite is positioned within, but independent of,

(a) (b)

*Fig. 2-4. Pilot tube guidance system consists of (a) an LED-illuminated target housed in the steering head, and (b) a camera-mounted theodolite (photos courtesy of National Clay Pipe Institute)*

*Fig. 2-5. The image from the camera-mounted theodolite transmitted to a digital display for the operator's use (photo courtesy of Akkerman)*

this assembly. The control panel, digital display, and steering deck enable the operator to control the advance of the pilot tube.

The installation process is repeated with the addition of more pilot tube sections. The operator should be constantly checking, verifying, and

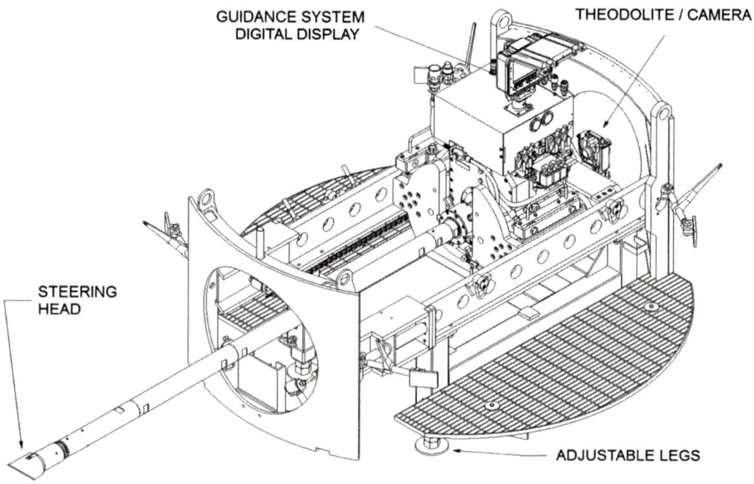


*Fig. 2-6. Typical jacking frame configured for pilot tube installation*

*Fig. 2-7. Typical jacking frame and pilot tube setup (photo courtesy of National Clay Pipe Institute)*

making corrections to steer the pilot tube through the ground with the center LED and crosshairs aligned. If the lights of the target begin to disappear from view, the operator is able to stop, retract the pilot tube, and try to reestablish visible confirmation of the light source, and then rotate the slant head to maintain the design line and grade of the pilot tube.

The operator is tasked with maintaining visible contact with the light source as the pilot tube advances. Equipment manufacturers use computer programs to enhance the images seen by the theodolite, which can increase the installation length possible.

Target (light source) visibility and clarity are affected by atmospheric conditions. These conditions might change during the course of the pilot tube insertion. Contributing factors might include

- Air quality, including dust;
- Differential of the temperature of the ground, pilot tubes, and air within the pilot tubes;
- Ambient air temperature;
- Humidity; and
- Heat from friction with the ground as the pilot tube string is inserted.

If some or all of these factors are present, a condition referred to as *fog* may occur inside the tubes. There are a number of methods to improve the atmosphere within the pilot tube, such as ventilation, introduction of inert gases, or simply waiting until the fog clears. If the operator at the end of the installation of the pilot tube is able to maintain visual contact with the light source, the pilot tube will be installed to the line and grade of the original setup of the jacking frame and the theodolite. Despite best efforts, sometimes it might be necessary to continue to push the pilot tube, even after losing visual contact with the light source, in an attempt to reach the receiving shaft. Completing the pilot tube installation in this manner carries a greater risk of deviating from the planned alignment.

When the steering head of the pilot tube reaches the receiving shaft, the pilot tube installation is complete. Assuming visual contact was maintained throughout the installation, the contractor now has two fixed points in space (the entry point at the jacking shaft and the exit point at the receiving shaft) to survey and confirm the design line and grade. However, if visual contact was not maintained throughout the installation, there may be a deviation between these two points, even when the two points match the design line and grade. If the installation is unsatisfactory, the pilot tubes can be retracted and reinstalled before starting the next step of the installation process. However, the retraction of the pilot tube is not always necessary; some minor misalignments can be corrected through the subsequent steps of the pilot tube path enlarging process. Also, some ground conditions make retraction and steering corrections difficult. The initial borehole may collapse with the retraction of the pilot tube, and, upon reinsertion, the pilot tube may tend to follow the previous borehole (the path of least resistance).

## 2.3 ENLARGING THE PILOT TUBE PATH

Once the pilot tube is installed and the line and grade checked, the pilot tube becomes the guide for all future enlargement steps. The number of steps varies, based on

- Final diameter and material type of the pipeline;
- Equipment available to the contractor;
- Ground conditions; and
- Contractor preference.

The enlargement is traditionally completed using one or more upsizing tools in one or more passes, or steps, coupled with a spoil removal system. Together with the pilot tube installation, these are commonly referred to as a two-step or three-step process. On the second step, while using the installed pilot tubes as a guide, these systems excavate the ground and transport the spoils back to the jacking shaft. If a third step of enlargement is used, the spoils can be sent forward to the receiving shaft or backwards to the jacking shaft.

The upsizing tools and spoil removal systems may include any or all of the following:

- Reamer heads or cutter heads;
- Thrust casings;
- Powered reamer or powered cutter heads;
- Thrust casings within the jacking pipe;
- Augers within a permanent casing; and
- Simple expander (mandrel).

The ultimate goal is to enlarge the borehole created by the pilot tube to a diameter that will allow the insertion of the jacking pipe. The upsizing tool is attached to the pilot tube sections. As the enlarging tool is advanced to excavate and remove the ground, the pilot tubes are systematically advanced and retrieved at the receiving shaft. The pilot tubes are placed back into the empty racks (Fig. 2-2) for use on the next drive or project. The equipment used to enlarge the pilot tube is centered on the pilot path. The enlargement process excavates the ground 360° around the pilot path. Therefore, unlike most utility pipe installations that rely on invert elevations, the surveying and construction layout for the pilot tube method relies on the elevation of the pipe centerline. During installation, the centerline of the pilot tube should coincide with the centerline of the final carrier pipe that has been calculated from the elevations called out on the plans.

Presented next are the different approaches (steps) and equipment used to enlarge the pilot tube path for insertion of the final jacking pipe.

### 2.3.1 Reamer Heads or Cutter Heads

The initial upsizing tool is attached securely to the end of the pilot tube string. This upsizing tool can either be a reamer head (Fig. 2-8) or a cutter head (Fig. 2-9), depending on ground conditions. Thrust casings are connected to the rear of this upsizing tool and systematically added as the system advances through the ground. Augers inside of the upsizing tool and thrust casings rotate in a manner to convey the excavated ground away from the upsizing tool and rearward into the jacking shaft. These upsizing tools are sized to match the thrust casings to which they are attached.

### 2.3.2 Thrust Casings

Thrust casings are used to advance the selected reamer head or cutter head and to convey the spoils to the jacking shaft. The thrust casings are jointed steel casing with an internal auger of equal length (Fig. 2-10). The thrust casing can be manufactured with different lengths and diameters, but

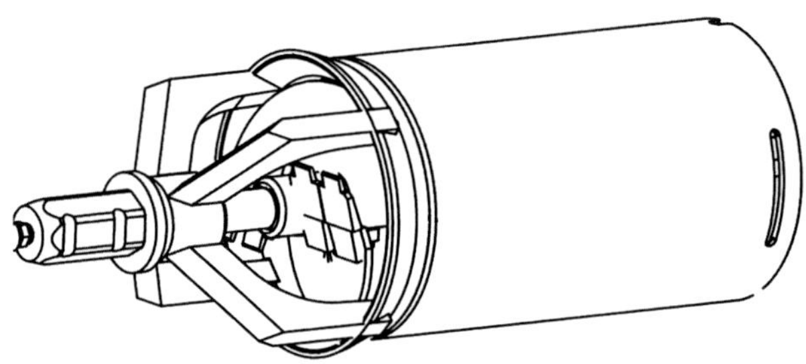

*Fig. 2-8. Schematic view of a reamer head assembly*

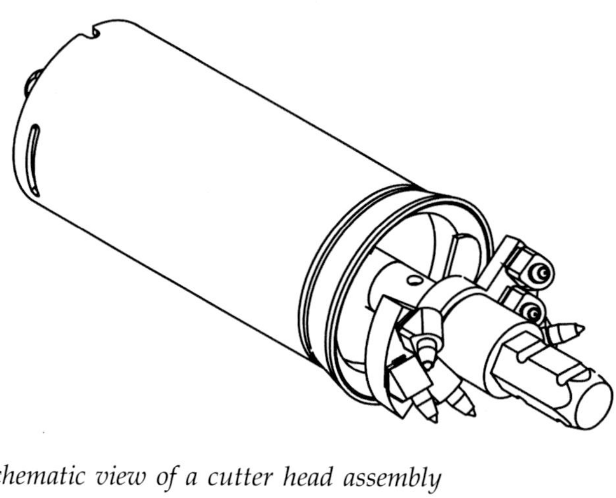

*Fig. 2-9. Schematic view of a cutter head assembly*

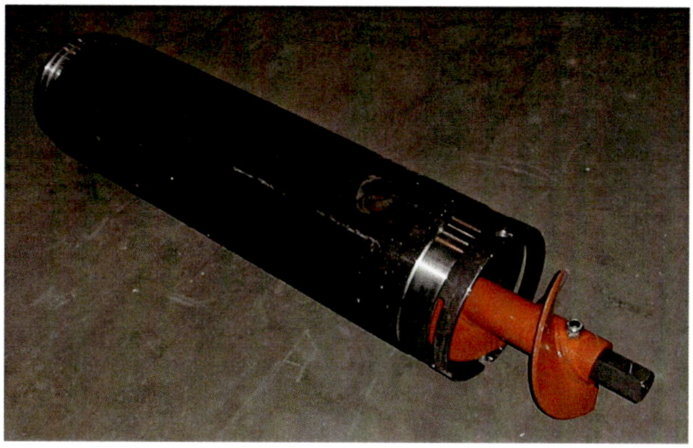

*Fig. 2-10. Thrust casing with auger used to transport spoils (photo courtesy of Akkerman)*

common thrust casing lengths are 39 in. (1.0 m) or 60 in. (1.5 m), and common diameters are 11 in. (275 mm) or 16 in. (405 mm). Fig. 2-11 shows a rack of thrust casing being used on a typical project. Like the pilot tubes, the thrust casings are added at the jacking shaft. As the thrust casings are jacked through the ground, they transport the spoils to the jacking shaft while advancing the reamer head or cutter head (Fig. 2-12). The spoils are collected into a skip pan and removed from the jacking shaft.

*Fig. 2-11. Rack of thrust casings used to displace the pilot tubes and enlarge the initial hole (photo courtesy of Akkerman)*

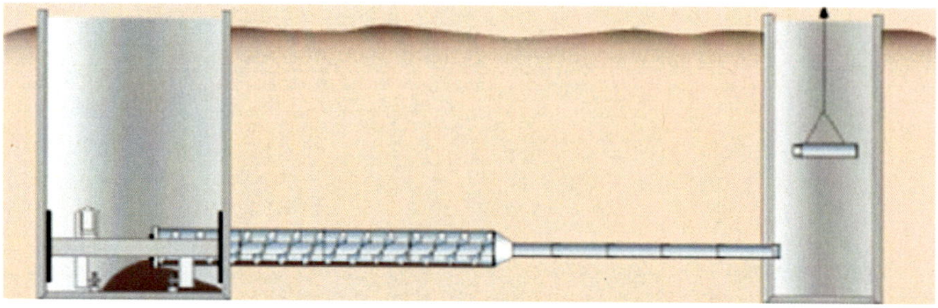

*Fig. 2-12. Installing thrust casings to enlarge the pilot tube hole by excavating the ground (photo courtesy of Bohrtec)*

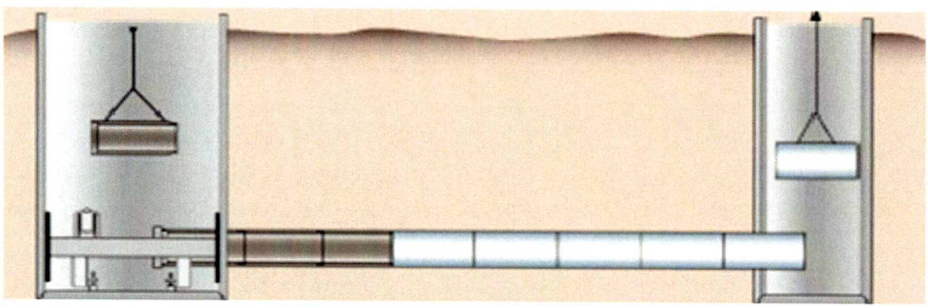

*Fig. 2-13. Installing the final carrier pipe as the thrust casings are advanced into the receiving shaft (photo courtesy of Bohrtec)*

*Fig. 2-14. Installing the final carrier pipe (photo courtesy of National Clay Pipe Institute)*

As the thrust casings are installed, the pilot tubes are advanced into, and recovered from, the receiving shaft. Once the thrust casings are installed for the full length of the drive, they will be advanced into (and removed from) the receiving shaft as the carrier pipe is jacked into the excavated hole from the jacking shaft (Figs. 2-13 and 2-14). Like the pilot tubes, the thrust casings are collected at the receiving shaft as they are advanced into the receiving shaft. In this instance, the OD of the final carrier pipe matches the OD of the thrust casings.

### 2.3.3 Powered Reamer Heads or Powered Cutter Heads

For installation of pipe with an OD larger than the installed thrust casing, contractors can use a powered reamer head (PRH) or powered cutter head (PCH) attached to the previously installed thrust casing (Figs. 2-15 and 2-16). The PRH and PCH are manufactured in different sizes and can be skinned to accommodate various diameters of jacking pipe, as discussed in Chapter 6. Housed within these powered heads are one or more hydraulic motors, depending on the type of head. The powered head is advanced by adding and pushing the jacking pipe forward.

The hydraulic motors of the PRH and PCH turn the auger flights within the thrust casing in the reverse direction to transport spoils forward to the receiving shaft (Fig. 2-17). The PCH has a rotating cutter face with openings to allow the spoils to enter. The PRH does not rotate, and the spoils enter as the PRH advances. The contractor's choice of whether to use a PRH or a PCH is based on the consistency and strength of ground, jacking pipe diameter, prior experience, and machine capabilities.

*Fig. 2-15. Typical powered reamer head assemblies displayed with 11-in. thrust casing and various size pipes (photo courtesy of National Clay Pipe Institute)*

*Fig. 2-16. Powered cutter head is added to the thrust casings to install even larger diameter carrier pipe. Demo shows from right to left: pilot tube, 16-in. reamer, 16-in. thrust casing, 36-in. powered cutter head, jacking pipe, jacking frame, and hydraulic power unit (photo courtesy of National Clay Pipe Institute)*

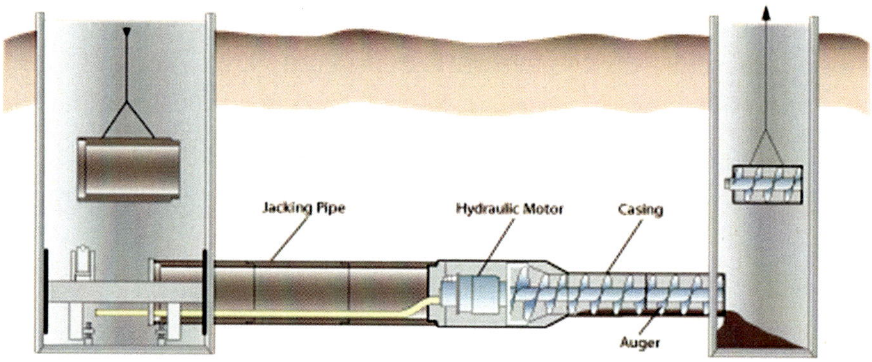


*Fig. 2-17. Upsizing with a powered head to install a jacking pipe (photo courtesy of Bohrtec)*

Although not common practice, both the PRH and PCH can be broken down into two sections to be launched and removed within small jacking and receiving shafts. Hoses are run through the jacking pipe to provide the hydraulic pressure and flow rate necessary to drive both of these powered heads. Customarily the jacking pipe is staged on the surface with the hoses installed before the launch of the PRH or PCH to reduce installation time. These powered heads are designed to be attached to the thrust casings.

As the jacking pipe is added at the jacking shaft, the jacking frame thrusts the jacking pipe and PRH/PCH forward, advancing the thrust casings into the receiving shaft for removal. Along with the hydraulic hoses, lubrication lines are installed in the jacking pipe to lubricate the overcut at the PRH/PCH as the jacking pipe is advanced. If the pipe diameter is large enough for personnel entry, lubricant can also be added to the overcut through ports within the jacking pipe. Jetting lines may be installed as well to lubricate the excavated spoils to help auger transport through the thrust casings. The augers in the thrust casing are rotated by the powered head and the spoils are transported to the receiving shaft for subsequent collection and removal. Like the pilot tubes, the thrust casings are advanced and collected in the receiving shaft for reuse on future drives. At the end of the installation, the powered head is also removed from the receiving shaft, leaving the jacking pipe in the bored hole.

### 2.3.4 Thrust Casings within a Jacking Pipe

Another enlargement method used by contractors is a simple auger housed inside its own small-diameter casing, which in turn is housed inside of the finished jacking pipe. With the development of the powered upsizing tools, this method is now seldom used. A reamer head or cutter head with a reduced-diameter internal auger is attached to the pilot tube string (Fig. 2-18).

These reamer and cutter heads are manufactured to different diameters to match the OD of the jacking pipes (Fig. 2-19). The advantage to contractors is that they are able to install multiple jacking pipe sizes while using only one size of thrust casing. This is facilitated by using centralizers to position the auger casing within the finished jacking carrier pipe. During installation of the jacking pipes, the pilot tubes are recovered in the receiving shaft while the spoils are simultaneously transported through the thrust casings back to the jacking shaft (Fig. 2-20).

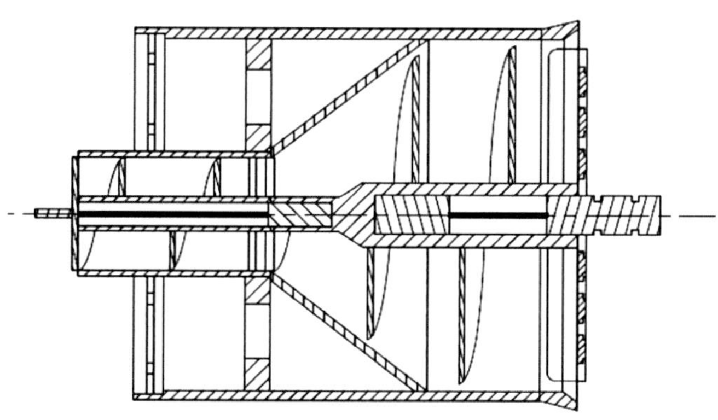

*Fig. 2-18. Cross section of a cutter head with a reduced-diameter internal auger*

*Fig. 2-19. Cutter head (a) front view with the leading edge cutting wheel and (b) rear view with transition to the trailing auger and casing and lubrication tubing to lubrication ports (photos courtesy of National Clay Pipe Institute)*

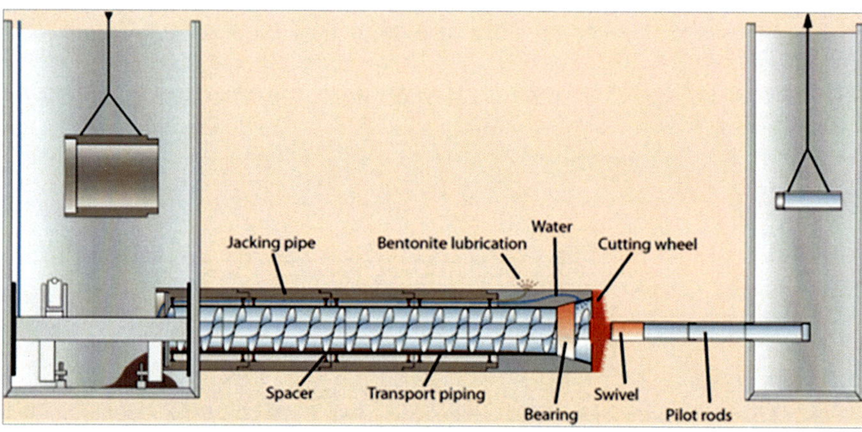


*Fig. 2-20. Typical thrust casing within a jacking pipe (photo courtesy of Bohrtec)*

### 2.3.5 Augers within a Permanent Casing

Another enlargement method is to use augers directly inside a permanent casing, much like an auger bore setup. This method is typically associated with small-diameter [24 in. (610 mm) and less] installations with short casing lengths to work from within the small-diameter jacking shafts. The pilot tube transitions to a piece of casing with centralizers to keep the casing centered on the pilot tube's initial hole (Fig. 2-21).

*Fig. 2-21. Transition from the pilot tube to the permanent casing (photo courtesy of National Clay Pipe Institute)*

The forward progress of the augers is bounded by the centralizers at the leading edge of the casing. As the casing is advanced, the centralizers help to slice the ground. The ground enters the casing, forming a temporary earth plug. The plug provides partial face stability during casing advancement. The augers inside of the casing are then used to convey the spoils from the leading edge of the casing back to the jacking shaft. This method only requires the use of the pilot tubes and the manufactured transition casing. Both items are recovered from the receiving shaft as new casing pieces are added to the pipe string. The final step is to clean out the casing with the augers. For this enlargement method, a final carrier pipe is usually installed within the casing.

This same enlargement method is routinely used with large-diameter [72 in. (1.83 m) and less] installations when a horizontal auger boring rig is used. See the discussion in Section 2.4.1.

### 2.3.6 Simple Expanders (Mandrel)

For small-diameter pipe installations [up to 8 in. (200 mm)], it may be appropriate to use a simple expander to enlarge the hole by physically displacing the adjacent soils. This method of hole enlargement does not involve the removal of soil cuttings. With this approach, after the initial pilot tube installation is completed, the expander is attached to the pilot tube rods, which are then used to push or pull the expander along and through the pilot tube alignment. The diameter of the expander is slightly larger to create an overcut for the trailing carrier pipe. Examples of hole expanders are shown in Fig. 2-22. In some cases, the carrier pipe may be located immediately behind the expander (Fig. 2-23) and pushed or pulled into the final hole.

(a) (b)

*Fig. 2-22. Examples of simple expanders attached to the pilot tube to allow a pipe string to be pulled back through the initial pilot tube hole. The expander enlarges the initial hole during pull-back (photos courtesy of Nick Strater)*

*Fig. 2-23. Bundled conduits attached to a simple expander during pull-back (photo courtesy of Nick Strater)*

## 2.4 PILOT TUBE AS A GUIDE FOR OTHER TRENCHLESS METHODS

As mentioned previously, once the pilot tube is inserted into the ground, it becomes the guidance path for the enlargement of the initial hole and installation of the final carrier pipe. The enlargement process can be completed by a number of other trenchless methods. Examples of the different methods are presented next.

### 2.4.1 Horizontal Auger Boring

Traditional horizontal auger boring has limited accuracy because little technology exists to determine the position of the leading edge of the casing. The grade at the leading edge of the casing is usually monitored with a water level. Both the line and grade can be checked by removal of the augers and personnel entry into the casing to conduct a survey. To accomplish this, casing diameter and face (ground) stability must be suitable for personnel entry. Further details regarding horizontal auger boring can be found in *ASCE Manual of Practice (MOP) No. 106* (2017).

In the last decade, the pilot tube method for installing pipes on line and on grade has had an impact on the horizontal auger boring market, improving the accuracy of this method (Figs. 2-24 and 2-25). The method is referred to as the guided boring method.

Once the initial guided pilot bore is completed, a conventional horizontal auger boring rig is attached to the pilot tube (Fig. 2-26). The horizontal auger boring technique can then be used. It follows the alignment of the pilot tube by the use of an upsizing adapter, allowing for greater accuracy of the horizontal auger bore. When installing larger-diameter casing, it is common to use an intermediate step and install thrust casings and then adapt to the horizontal auger boring rig for final casing installation. This reduces the stresses on the connections, thereby reducing risk of failure. The final casing diameter can often be reduced because of the accuracy of the installation.

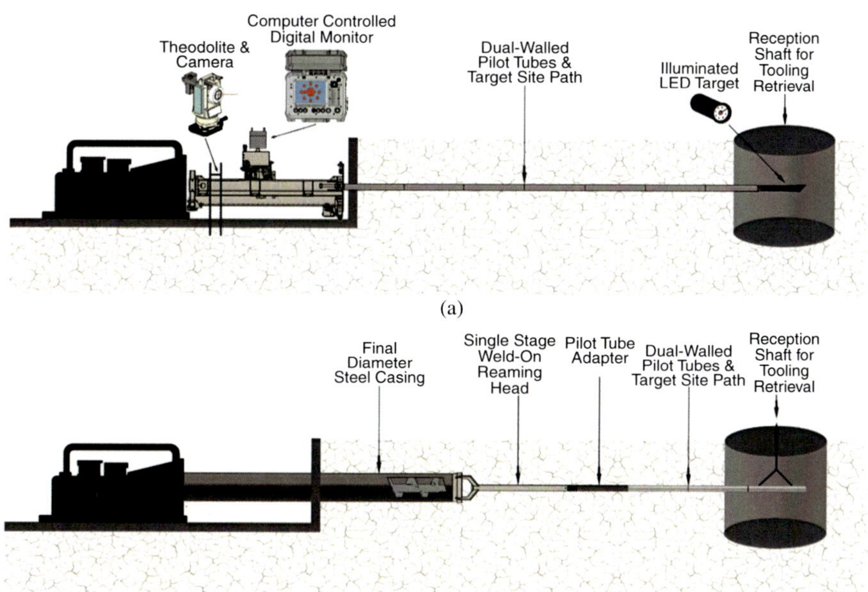


*Fig. 2-24. (a) The horizontal auger boring rig is set up in the jacking shaft with guide rails. The pilot tube frame is set in front of the rig, and a pilot tube is inserted into the ground; (b) After the pilot tube is installed, an upsizing adapter is attached to tie the horizontal auger boring rig to the pilot tube for guidance*

*Fig. 2-25. Pilot tube frame setup in front of the horizontal auger boring rig to install the pilot tube for guidance (photo courtesy of National Clay Pipe Institute)*

*Fig. 2-26. An upsizing adapter is connected from the pilot tube to the horizontal auger boring rig to install a larger-diameter casing (photo courtesy of National Clay Pipe Institute)*

### 2.4.2 Pipe Ramming

Pipe ramming is a technique for installing a steel casing into the ground from a launch shaft by means of ramming, or pushing, the pipe through the ground using a pneumatic hammer or rammer, or a hydraulic jacking system. The dynamic forces and energy transmitted by the percussion hammer, which is attached to the end of the steel casing, drive the casing through the ground.

The orientation of the casing is typically controlled by guide rails set up in the launch shaft. As it is driven into the ground, the casing goes in the direction it is aimed by the guide rails. Sometimes, depending on the in situ density of the ground, the casing may begin to deflect from the planned line and grade in any direction. Historically, pipe ramming has little to no ability to steer. The casing grade can be monitored, but the casing goes where it is aimed.

In the last decade, the pilot tube method for installing pipes on line and on grade has had an effect on the pipe ramming market, improving the accuracy of this method (Fig. 2-27).

Once the initial guided pilot bore is completed, open-ended upsizing adapters are attached to the pilot tube (Fig. 2-28). A number of upsizing adapters will be used in series to go from the pilot tubes to slightly larger casing diameters, and then to the final casing diameter. Multiple adapters are needed to help reduce the shearing forces during the transition from a small-diameter pilot tube to the final casing diameter because the combined mass of the rammed casing and spoils is much heavier and will tend to follow its own path through the ground. By progressively upsizing, this tendency can be mitigated. The ability to install a rammed pipeline at a precise line and grade has improved by using a pilot tube for guidance.

### 2.4.3 HDD

Another use for the pilot tube method is to improve the accuracy of HDD installations. There are two principal approaches to this method.

*Fig. 2-27. Pilot tube installed as a guide for the pipe rammed casing (photo courtesy of Claude H. Nix Construction)*

*Fig. 2-28. Step-ups in casing size are typically used to transition from the pilot tube to the final casing diameter to help transfer shear from the small diameter to the large diameter (photo courtesy of Akkerman)*

In the first, after the pilot tube is installed between the jacking shaft and the receiving shaft, the pilot tube jacking frame is removed from the jacking shaft and replaced with an HDD rig. The HDD rig is used to enlarge the pilot tube borehole and, ultimately, to install the carrier pipe or casing from within the jacking or receiving shaft (Fig. 2-29).

*Fig. 2-29. The pilot tube is installed to line and grade and the borehole is then enlarged by the HDD rig as the pipe is pulled back (photos courtesy of Michelle Macauley)*

The second approach is for the HDD rig to be positioned on the surface and the drill rods advanced with a geometry that will allow them to pass through the jacking and receiving shafts. Upon reaching the receiving shaft, the drill rods are connected to the pilot tubes with an adapter, and the pilot tubes are used to guide the installation of the on-grade section of the pipe.

In both cases, the pilot tubes are typically pushed into the receiving shaft, where they are disassembled and removed while HDD drill rods are added and installed.

Once all of the HDD drill rods are in place, the HDD rig can be used to pull a reamer and the final carrier pipe or cable back through the borehole using traditional HDD techniques. The advantage of using the HDD rig for final pipe installation is its pulling capacity, which often exceeds the pulling capacity of the pilot tube system.

If a reamer is used with the HDD equipment to enlarge the borehole, bentonite-based drill fluid might be required to remove cuttings from the bore annulus. The contractor will need to collect and recycle this drill fluid from one or both of the shafts.

In these applications of the pilot tube method, the carrier pipe is typically high-density polyethylene (HDPE) or fusible or mechanically coupled PVC. It can be assembled and pulled from the surface as is typical for HDD, or it can be installed working from within one of the shafts. If working from within a shaft, smaller-diameter pipes can be pulled from a coil located either on the surface or on the shaft bottom. Larger-diameter pipes that cannot be coiled or cannot manage the bends required to be pulled in from the surface must be assembled in the shaft bottom from sticks of straight pipe (Fig. 2-30).

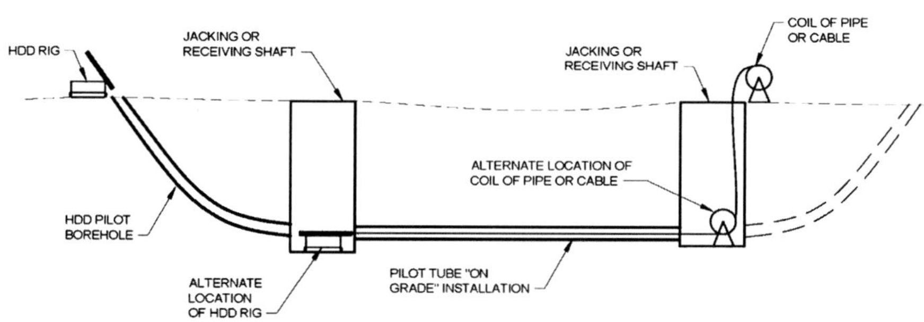


*Fig. 2-30. The HDD rig can also be set up on the surface and drilled down to the pipe horizon where the pilot tube is used to install to line and grade the borehole at the bottom of the invert HDD profile between the jacking and receiving shafts*

## 2.5 VARIATIONS AND REFINEMENTS TO THE PILOT TUBE METHOD AND OTHER GUIDED BORING METHODS

The basic concept of the pilot tube method is to use a hollow tube to transmit a light source for guidance control. Traditionally, the method has been applied in displaceable soils in which augers rotating within a casing convey spoils in subsequent steps. Recent equipment innovations have allowed the installation of pilot tubes in nondisplaceable soils and rock, and they use alternative methods to transport spoils. The following sections describe a number of modifications and adaptations currently available.

### 2.5.1 Pilot Tube Steering Head for Nondisplaceable Soils and Rock

Installing pilot tubes on line and on grade in nondisplaceable soils and rock is now possible with a drill bit adapted to work with the pilot tubes. One system, developed by Akkerman (2016), is called Rock Drill Adapter (or RDA). The process involves a hard ground steering head that uses commercially available rock drill bits to install the pilot tubes to a high degree of accuracy in nondisplaceable soil or rock. There are multiple types of drill bits that can be used for different strengths of nondisplaceable soil or rock. The technology works best in somewhat homogeneous geologic conditions in which the bit can engage an entire hard ground or rock mass. The drill bit is not intended to work through cobbles and boulders, in which the drill bit is likely to be deflected off of its intended path.

During this type of pilot tube installation, the steering head actually drills into the ground and generates cuttings as opposed to merely displacing the ground as described in Section 2.2. The cuttings are then transported back to the jacking shaft through the annular space around the outside of the pilot tube created by the overcut of the drill bit. A drilling fluid is pumped through a fluid channel of the dual-walled pilot tube (Fig. 2-31) to the steering head where it exits through ports in the drill bit (Fig. 2-32). The cuttings and drilling fluid are collected at the jacking shaft for disposal. The drilling fluid is used to:

- Cool the drill bit;
- Flush the cuttings away from the drill bit; and
- Transport the cuttings back to the jacking shaft.

Guidance of the drill bit and pilot tube is achieved in the same manner as outlined in Section 2.2. Steering of the pilot tube in nondisplaceable soil or rock is achieved by using an angled steering head similar to the one used for displaceable soils. The key difference is the use of tungsten carbide cutting tools on the drill bit to aid in the cutting (Fig. 2-32). The steering principle is fundamentally the same in that the operator uses the illuminated indicator

*Fig. 2-31. Dual-walled pilot tube used with a hard ground steering head (photo courtesy of Akkerman)*

*Fig. 2-32. Various steering heads available for the pilot tubes in nondisplaceable or rock conditions (photo courtesy of Akkerman)*

on the target to orient the steering head in the direction that he or she wants the pilot tube to go. With the cutting tools placed along the crown and leading edge of the drill bit, a carving method is used to etch a relief in the ground in the direction of desired movement to maintain the alignment. Because of the increased difficulty associated with drilling in rock compared with the traditional soil displacement method, the operator is encouraged to hold a very tight tolerance on the alignment so that only small corrections are needed to maintain the desired line and grade.

A critical element to the successful installation of a pilot tube in hard soils or rock is maintaining the flow of cuttings back to the jacking shaft through the annulus. If this flow is interrupted, increased thrust and/or rotational forces, or seizing of the pilot tube can occur as a result of not being able to clear the cuttings from the drill bit and/or annular space. Because each ground condition is unique, different drilling fluid mixes may be required.

### 2.5.2 Guided Boring without a Pilot Tube

A number of manufacturers have developed, and continue to develop, guided boring systems that allow installations in nondisplaceable ground and rock without the need for a pilot tube. These methods all use slightly different proprietary techniques but typically replace the pilot tube installation step with the use of a steerable boring head, similar in concept to microtunneling, tunnel boring machine pipe jacking, and small boring units, at the leading end of segmented temporary steel casings (similar to the thrust casings of traditional pilot tube methods). Most of the systems use hollow-stem augers rotating in thrust casings to remove the spoils from the cutter head. The hollow center in the augers serves as the optical path for guidance, similar to a pilot tube. These systems are designed for use in nondisplaceable ground and are considered marginal for use in displaceable ground. Therefore, installations in displaceable ground should be approached with caution. The following paragraphs describe some of the systems available at the time of writing.

**2.5.2.1 Front Steer.** The Front Steer (patented) system, developed by Bohrtec, uses an optical guidance system consisting of a theodolite, closed-circuit television (CCTV) camera, monitor, and LED target board. As shown in Fig. 2-33, the ground is continuously excavated by the cutter head and then transported to the jacking shaft by hollow-stem augers. The machine operator can perform steering movements remotely by articulating the Front Steer unit. At the same time, the control panel shows the respective steering position of the head (Fig. 2-34). The cutter head ranges from 16 in. (405 mm) to 36 in. (915 mm) in diameter and uses standard thrust casings for advancement and spoils transport. According to the manufacturer, Front Steer enables the use of guided boring technology in soil with SPT N-values greater than 35 and in rock with strengths of up to 2,900 psi (20 MPa).

**2.5.2.2 Eliminator.** The Eliminator system (patented), developed by Akkerman, uses hollow-stem augers and is optically guided using typical pilot tube guidance system components with an LED target. Hollow-stem augers are used to drive the cutter head and transport cuttings back to the

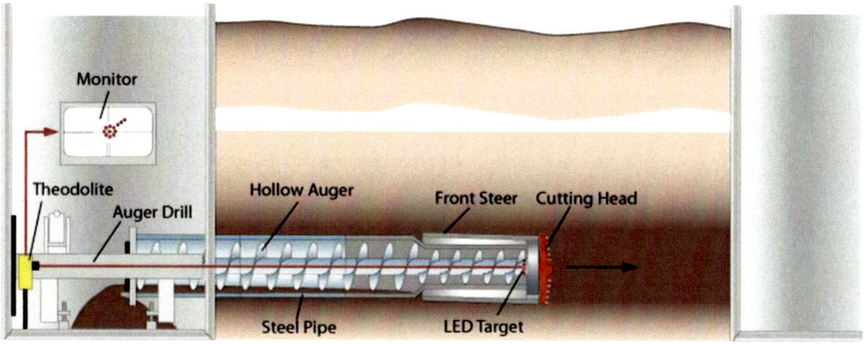


*Fig. 2-33. Cross section of a Front Steer system in operation (photo courtesy of Bohrtec)*

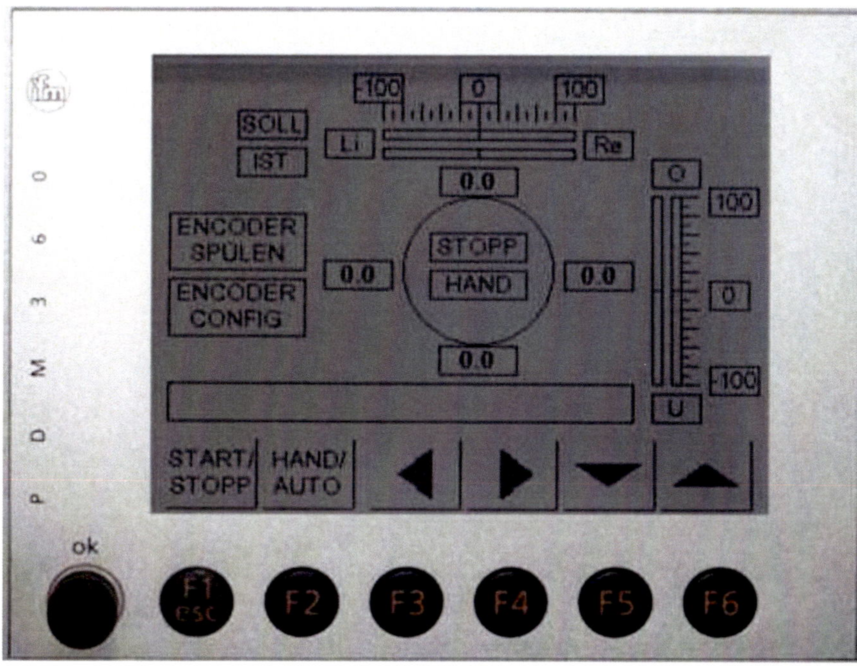


*Fig. 2-34. Control panel for the Front Steer system (photo courtesy of Bohrtec)*

jacking shaft through the thrust casings (Fig. 2-35). Several cutter head configurations are available and can be tailored for specific ground conditions. Steering is provided by three independently controlled, hydraulic shoes. The steering head is equipped with high-pressure jetting nozzles to

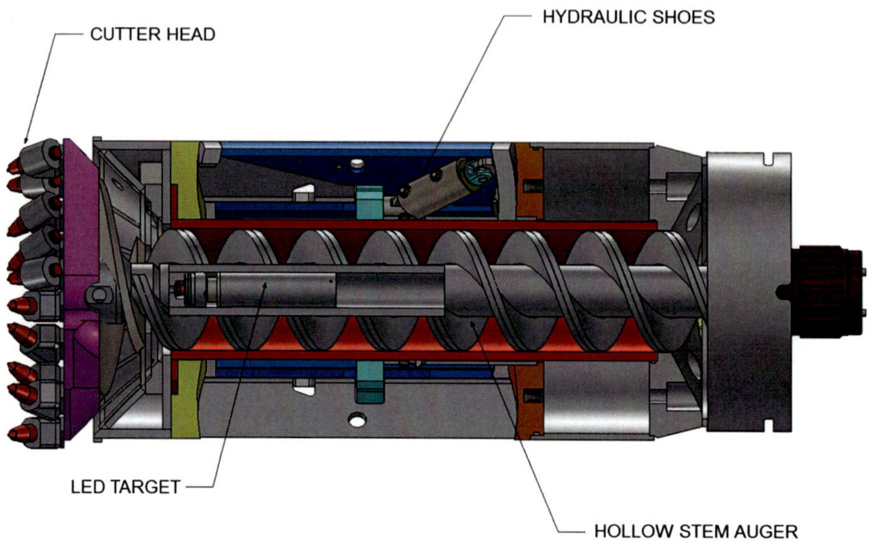


*Fig. 2-35. Cross section of an Eliminator system*

aid with cutting or lubrication during operation. Once launched, the Eliminator can be retrieved by pulling it back to the jacking shaft. This functionality is extremely beneficial in the event that there is a change in ground conditions, allowing the operator in such instances to make a cutter bit change.

The cutter head is 16 in. (405 mm) in diameter and uses standard thrust casings for advancement and spoils transport. Once the thrust casings are installed, additional upsizing tools can be used to increase the borehole diameter. As discussed in Section 2.3.3, jacking pipe can be installed from the jacking shaft, advancing the thrust casings into the receiving shaft. According to the manufacturer, the Eliminator enables the use of guided boring technology in soil with SPT N-values greater than 30 and in rock with strengths of up to 10,000 psi (70 MPa).

### 2.5.3 Using a Vacuum Excavator to Remove Spoils Instead of an Auger

The AXIS (patented) guided boring system, by Vermeer Corporation, uses an articulating cutter head (Fig. 2-36) and vacuum system to excavate and remove the spoils while guidance is provided using a laser light source and reference target communicating through a sight channel (Fig. 2-37).

The AXIS system is designed to install 10-in. (254-mm) to 18-in. (457-mm) pipe at typical lengths up to 350 ft (107 m). A hydraulic motor on the jacking frame rotates the cutter head through a central drive shaft. A vacuum/suction line is applied to one of the pathways within the thrust casing to transport cuttings back to the jacking shaft.

*Fig. 2-36. Vermeer's AXIS system with an articulating head for steering control (photo courtesy of Vermeer)*

*Fig. 2-37. Rack of AXIS thrust casings. Each casing has a central drive shaft and openings for the light source guidance system, power cables, and a return for the vacuum excavator (photo courtesy of Norm Joyal)*

Guidance for the AXIS system is provided by a pipe laser, which is set in the jacking shaft at the alignment and grade the project requires. Contained within the drill head is a camera connected to a monitor on the operator console that relays the position of the laser at the drill head. The excavated

*Fig. 2-38. Typical setup of the AXIS system with the vacuum line attached to the thrust casings (photo courtesy of Norm Joyal)*

ground is vacuumed out through a hollow section in the drill head and subsequent drill casing to a vacuum tank at the jacking shaft (Fig. 2-38). The process is repeated, adding sections of drill casing until the cutter head reaches the receiving shaft. Next, the cutter head is removed at the receiving shaft and then either the carrier pipe is attached to the lead drill casing and pulled in, or, if jacking pipe is used, it can be advanced from the jacking shaft. In both cases, the temporary thrust casings are recovered into either shaft for recovery and reuse. The system's use of vacuum for spoils transport in noncohesive ground conditions must be approached with caution. Similarly, vacuum systems may be of marginal effectiveness in high-plasticity clays.

### 2.5.4 Other Guidance Systems Using Internal Light Sources

As outlined previously, guidance systems equipped with a theodolite/ camera (with a visible light source seen through the pilot tube or hollow auger stem) are used to accurately install pipelines. Some horizontal auger boring machine manufacturers have added the light sources to their steering heads. The lights shine back through a channel on top of the casing or through the casing itself to serve as a point of reference for the operator. One specific system is the On Target system (patented) by McLaughlin. Although horizontal auger boring itself is not considered a guided system (not to be confused with steerable), equipment manufacturers are mounting light sources to assist with line and grade control. Detailed discussion of these systems is outside the scope of this MOP. The reader should consult *MOP 106* for more discussion of guided and steerable horizontal auger boring equipment.

## 2.6 APPLICATIONS

The pilot tube and other guided boring methods can be used to install pipelines with accuracy into the ground without a trench. This accuracy is useful when installing pipe for gravity systems. Typical applications include the installation of

- Main line gravity systems;
- Pipeline crossing;
- Laterals from manholes;
- Side connecting pipelines; and
- Pilot tube as a probe.

### 2.6.1 Installation of Main Line Gravity Collection Systems

Main line sanitary sewer collection systems are typically installed as a linear pipeline project. The systems collect flows upstream and carry flow to downstream points, which include other collection systems, pump stations, and/or treatment plants. These systems can be installed using open-cut trenches or with trenchless methods. For access to the collection system, the owner will incorporate maintenance access points or manholes. The spacing of the manholes is typically 250 ft (76 m) or more, depending on the size of the pipeline. When the pipeline system is installed using trenchless methods, the manholes become ideal locations for jacking and receiving shafts. Fig. 2-39 shows a typical layout of alternating jacking and receiving shafts. Jacking shafts can be used to install pipe in multiple

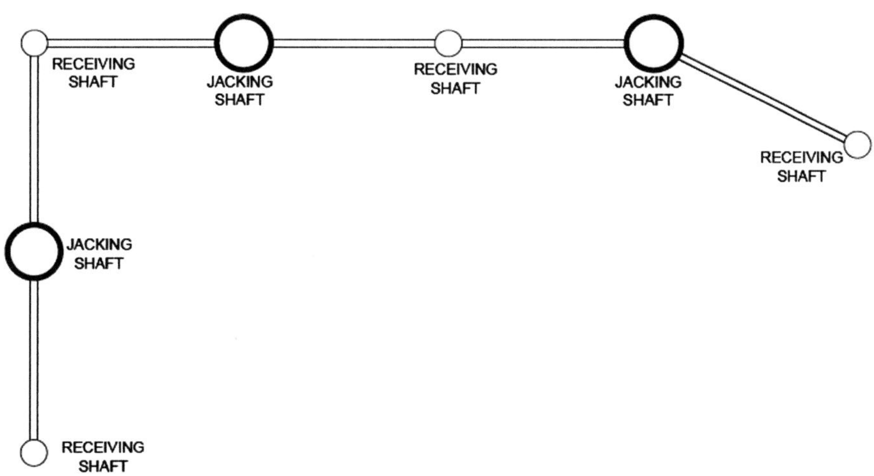


*Fig. 2-39. A typical layout of alternating jacking and receiving shafts used to install a linear gravity pipeline system*

directions. Receiving shafts are used to receive the trenchless equipment used to install the pipelines. Circular shafts can be used as turning points for the pipeline systems. The pilot tube method is ideally suited for installing pipeline systems because the method can work out of small-diameter shafts with typical drive lengths of 300 ft (91 m) and diameters associated with typical sanitary sewer collection systems. The reader is directed to Chapter 3 for more information on the planning of a pilot tube project.

### 2.6.2 Installation of a Pipeline Crossing

For many pipeline projects, a pipeline is installed by open cutting a trench, placing subbase foundation material, laying the pipe into the trench, and then backfilling the trench. These linear projects will encounter physical obstacles that will prevent the use of open-cut trenching. These physical obstacles include and are not limited to

- Highways,
- Railroads,
- Utilities,
- Walkways and bike paths,
- Parks and golf courses,
- Waterways (environmentally sensitive areas, creeks, streams, rivers), and
- Airport runways and taxiways.

To cross these obstacles, a trenchless construction method is typically used (Fig. 2-40). Crossings typically range from 100 to 300 ft (30.5 to 91 m) in length, with diameters up to 48 in. (1.2 m); therefore, as outlined in Table 1-1, pilot tube and other guided boring methods can be suitable for

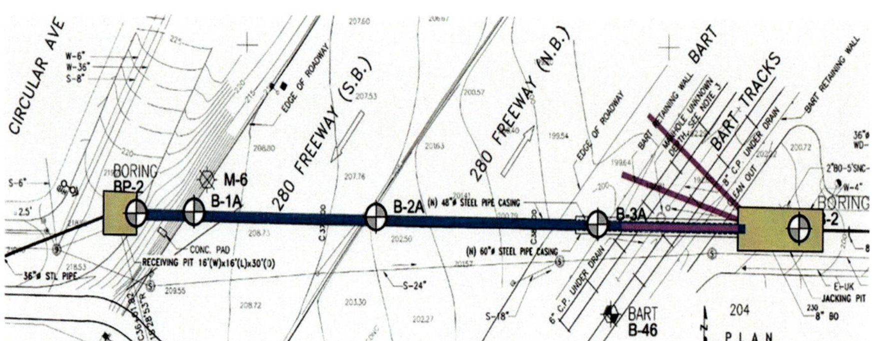


*Fig. 2-40. Example of a trenchless crossing under surface tracks of a major transit system (BART) and the northbound and southbound lanes of Interstate 280 (courtesy of Glenn Boyce)*

these applications. Other trenchless methods may be required when the crossings are longer, larger in diameter, or below the groundwater table.

### 2.6.3 Installation of Laterals from Manholes

Historically, pipelines are laid in trenches. Service connections (drainage and sewer laterals) are made at any location along the pipeline because access is available during construction. However, pipelines with multiple service connections tend to allow the infiltration of groundwater over time and can be difficult to inspect and rehabilitate in the future.

In an effort to reduce the number of service connections between manholes, the pilot tube method was developed in Germany in the 1980s to install 4-in. (100 mm) to 6-in. (150 mm)-diameter sewer laterals from 8-ft (2.4 m)-diameter shafts. The main line sewers were installed by trenchless methods, and the jacking and receiving shafts were the only access points to the new pipeline. Service connections can be made in any direction and from any desired depth to the manhole. This radiating pattern of service connections from a manhole was first developed in Berlin, Germany, and is often referred to as the Berlin construction method (Fig. 2-41). By using the Berlin construction method, manholes become access points to the sewer laterals, allowing for simpler inspections and routine service. Inspection and rehabilitation of the main pipeline is also simpler because the service connections are made only at the manholes.

As was presented in Section 2.6.1, the pilot tube method can now be used to install the main line sewers between the shafts as well.

### 2.6.4 Installation of Side Connecting Pipelines

As explained in the previous subsection, the pilot tube method is ideally suited to install small-diameter pipelines horizontally from a vertical manhole to make house connections. The same concept can be used to install small-diameter pipelines inclined from a large-diameter sewer/tunnel/interceptor to make connections to an existing sewer or manhole.

This concept was used on a project in Sacramento, California, where a new interceptor was being constructed and a connection to an existing sewer was needed. The concept design is shown in Fig. 2-42. The contractor mobilized a pilot tube frame in the newly constructed tunnel and installed the connecting side sewer to an existing manhole (Fig. 2-43).

### 2.6.5 Use of the Pilot Tube as a Probe

The reliable line and grade accuracy associated with the pilot tube method makes it possible to install pipes in close tolerances to existing utilities. The pilot tubes installed in the first step in excavation have the

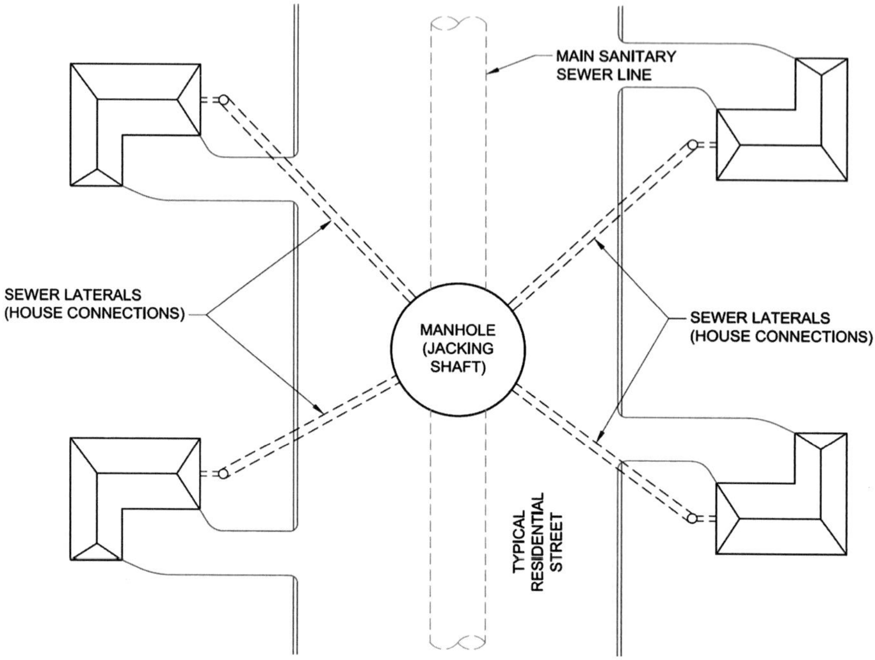


*Fig. 2-41. Typical manhole/jacking shaft configuration where pilot tube guided sewer laterals are installed to make house connections*

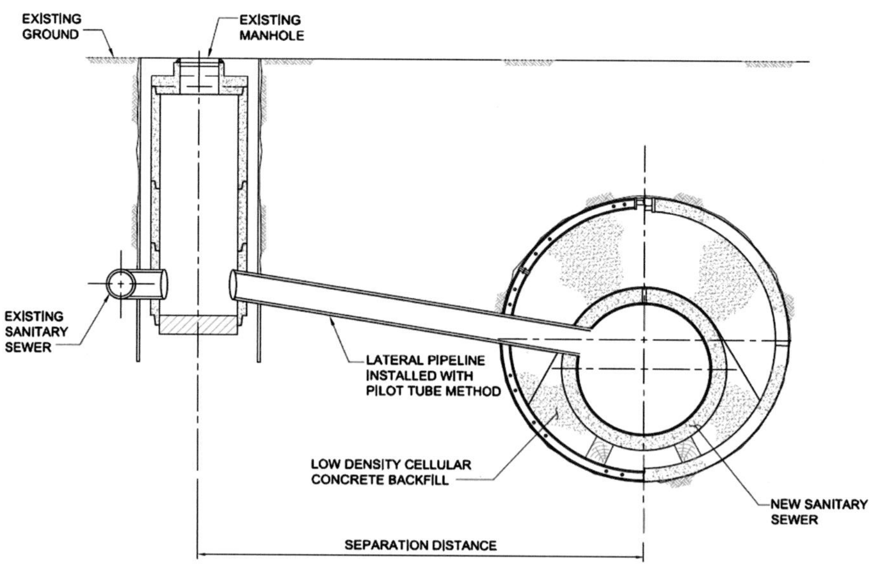


*Fig. 2-42. Layout for the construction of a side sewer connection from a newly tunneled interceptor to an existing manhole*

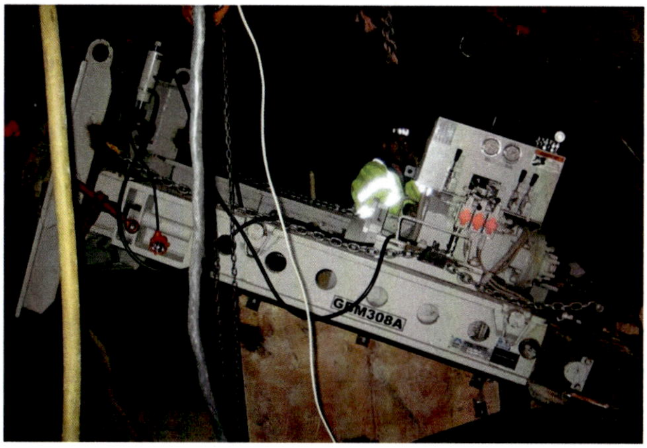

*Fig. 2-43. Pilot tube jacking frame angled to install an inclined connecting sewer from a new large-diameter tunneled interceptor to an existing manhole (photo courtesy of Andrew Finney)*

capability to discover unknown underground obstacles prior to making a full commitment to the entire bore, thereby eliminating an unplanned receiving shaft, which would be required in most other trenchless methods.

Pilot tubes have been installed in advance of other trenchless methods as a probing tool of the ground conditions to be encountered and to verify the design alignment is clear. Because the pilot tubes are fully retractable, they can be installed in displaceable ground conditions to determine the presence of the following items:

- Cobbles or boulders;
- Rock-like conditions;
- Steel, concrete, and timber piles;
- Steel sheet piling;
- Abandoned footings;
- Existing and abandoned utilities; and
- Known or unknown buried structures.

The pilot tube is launched on a prescribed line and grade, and if it hits a buried object, the distance to the object from the shaft is known. The approximate location and elevation of the buried object is established, and further investigations can be conducted. The alignment of the new pipeline can be modified based on the new information, or the buried object can be removed by open-cut excavation and the probe hole backfilled. Multiple pilot tube probes can be undertaken from the same shaft using the same pilot tubes to check for buried objects. The probes can be undertaken in whatever pattern is needed to determine whether any of the previously listed features are present.

The number of probes is a function of the diameter of the pipeline or casing to be installed, the size of the buried objects, and the number of the objects to be found. When the pilot tube probe is inserted, the ground is being displaced and, in part, densified. If the pilot tube probe is then retracted, the created hole may stand open infinitely, stand partially open with some hole raveling, or be returned to its original loose state. How the ground behaves when the pilot tube is retracted could affect the spacing of additional probes and how difficult or easy it is to install the next probe. If backfilling a pilot tube probe hole is required, it is important to consider whether the backfill will have a negative impact on the future trenchless installation.

## 2.7 BENEFITS AND LIMITATIONS

The pilot tube method has the following benefits:

- It can establish accurate on-grade alignments.
- It uses smaller shaft sizes, affording greater flexibility in project layout, and reduces surface impact.
- It can be used in soft to medium-stiff displaceable soils as well as in nondisplaceable soils and rock using the appropriate tooling.
- It can be used to install a variety of pipe materials.
- It can be used as a probe to provide valuable information about actual ground conditions.
- The pilot tubes can be retracted and reinserted or relocated to correct deviations in line and grade.
- The pilot tubes can be used to assist with guidance on other trenchless techniques.

The pilot tube method has the following limitations:

- The presence of gravels, cobbles, boulders, or buried objects can cause the pilot tube to deviate from line and grade or even stop the pilot tube.
- Shorter pipe sections result in more joints in the pipeline between shafts.
- It is limited in pipe diameter and drive length.
- It might not be suitable below the groundwater table, depending on ground conditions.

## REFERENCES

ASCE. (2017). "Horizontal auger boring projects." *Manuals and Rep. on Engineering Practice No. 106*, Reston, VA.

ASCE. (2015). "Standard design and construction guidelines for microtunneling." *ASCE 36-15*, Reston, VA.

Stein, D., Mollers, K., and Bielecki, R. (1989). *Microtunnelling: Installation and renewal of nonman-size supply and sewage lines by the trenchless construction method*, Ernst & Sohn, Berlin.

# CHAPTER 3
# PLANNING

## 3.1 INITIAL CRITERIA

Pilot tube and other guided boring methods are specialized construction methods suitable for conditions in which open-cut construction methods are not cost-effective, are too disruptive, or are not technically possible. As outlined in Chapters 1 and 2, pilot tube and other guided boring methods are typically used to install pipelines of 4 to 48 in. (100 mm to 1.2 m) in diameter, for distances of approximately 200–300 ft (61 to 91 m), and at or above the groundwater table. Installations below the groundwater may be possible and are limited by ground type, depth below the groundwater table, ground improvement methods, or specific equipment used. When the diameter of the pipeline increases, the distance between access points may decrease with using pilot tube and other guided boring methods. When work occurs below the groundwater table or larger diameters are to be installed, other trenchless methods may be needed to install the pipeline, such as:

- Horizontal directional drilling;
- Pipe jacking;
- Microtunneling;
- Utility tunneling; and
- Pipe ramming.

There are significant differences between these construction methods, both in application and in results. The engineer should review each of these methods to determine whether it can meet the project requirements and be used successfully under the specific conditions that will be encountered during the pipe installation process.

Careful evaluation of the following parameters will determine the applicability and constructability of a potential pilot tube and other guided boring project:

- Geotechnical conditions and geology, including groundwater;
- Potential for encountering objects (both natural and remnants of past construction) along the proposed alignment;
- Environmental conditions, including potential for contaminated ground or groundwater;
- Available work space for shafts;
- Required drive lengths;
- Pipe diameters;
- Cost and construction schedule;
- Alignment;
- Site access;
- Depth;
- Grade;
- Installation tolerances;
- Effects on surface activities;
- Location of existing, abandoned, or proposed utilities;
- Rights-of-way; and
- Location of existing, abandoned, or proposed structures.

The engineer of a pipeline project is encouraged to consider all trenchless methods before concluding that the pilot tube and/or other guided boring methods are the best construction methods available. Manuals and reports on engineering practice (known as *MOPs*) have been written by ASCE for different construction methods. A list of useful references is provided at the end of Chapter 1.

If the engineer responsible for the pipeline project does not have a strong background in trenchless design, an engineering firm that specializes in trenchless designs should be consulted to provide a peer review early in the planning/design process to help ensure good design choices are being made.

### 3.1.1 Pipe Diameter and Grade

The pipe diameter must be of sufficient size to meet the project requirements. For gravity sewers, the grade should be established to meet the project requirements and to accommodate the pipe diameter selected. Flow capacity (Q) and cleansing velocities must be considered for gravity sewers.

Design grades and tolerances are typically specified for gravity systems. Grades and tolerances are typically less critical for pressure pipelines and dry utilities and conduits. When grades are relatively flat or critical, or when unfavorable ground conditions exist, a two-pass approach should be considered, i.e., install a larger-diameter casing first and then install the carrier pipe inside the casing at the required line and grade.

The engineer should consider including at each manhole location an additional drop of 0.1 ft (30 mm) or more, plus the grade tolerance, to assist in maintaining the design grade during construction. Without this additional drop across the manholes, installation of the pilot tube could be within tolerance yet result in ponding within the manhole. Over the length of a project, these drops can be used to help maintain the overall grade in the pipeline if one section becomes misaligned during construction.

### 3.1.2 Depth

The following items should be considered when determining the depth of the pipe installed using pilot tube and other guided boring methods:

- Gravity flow or pressure pipeline;
- Required start and end points as well as intermediate pick-up points or tie-in points;
- Project performance (i.e., hydraulics, connections, operations, and maintenance);
- Geotechnical conditions and geology;
- Existing buried infrastructure;
- Future construction and changes to the ground surface elevation;
- Potential and tolerance for settlement or heave; and
- Potential and tolerance for inadvertent returns of lubricant.

The pilot tube and other guided boring methods are not as adversely affected by depth compared with open-trench construction. The major costs of deeper installations are shafts. Therefore, the elevations for pilot tube and other guided boring methods can be adjusted to take advantage of better ground conditions, to avoid other infrastructure, or to meet future needs when groundwater is not an issue. Adjusting depth is not always possible for gravity systems with flat slopes and when tying into an existing system.

Selection of minimum ground cover should take into consideration the ground conditions and use of the surface space above the pipe. Minimum ground cover is typically a function of gravity versus pressure flow, pipe diameter, and geology. As the ground cover decreases, the risk of surface settlement and heave increases. Ground treatment may be required to control surface settlement or heave with minimal ground cover.

Typically, a minimum cover is needed to allow sufficient depth for the ground to arch over the new pipeline installation. The minimum depth of cover is determined based on pipe outside diameter (OD), existing ground conditions, and other site conditions. Evaluation and calculation of potential settlement, heave, and inadvertent returns should be performed by the engineer to assist in the selection of the appropriate ground cover. Potential ground settlement or heave and the owner's tolerance to the attendant risk should be taken into consideration when establishing the depth and the amount of ground cover.

### 3.1.3 Routing

The routing of the pipeline can be affected by many factors, including:

- Required start and end points as well as intermediate pick-up points or tie-in points;
- Geotechnical conditions and geology;
- Surface topography;
- Existing utilities;
- Physical buried objects either placed naturally or the result of construction;
- Available rights-of-way;
- Maintenance of traffic during construction;
- Surface access; and
- Contaminated ground and groundwater.

Refer to *ASCE Manuals and Reports on Engineering Practice No. 46*, "Pipeline Route Selection for Rural and Cross-Country Pipelines" (1998), for more information.

Pilot tube has been used as a cost-effective alternative to, or in conjunction with, open-cut construction and other trenchless methods where undercrossings are required, surface disturbance is limited, trench support is challenging, conflicting utilities are encountered, or deep trenches are required.

Areas of contaminated ground or groundwater should be avoided if possible, especially at shaft locations. If they cannot be avoided, a trenchless installation is often a good choice because it minimizes the volume of material being excavated compared with open-cut trenching.

In urban areas, shafts and working spaces are typically located within a single lane of the main road and away from intersections. Additional working areas may be required for pipe storage and office trailers off the road right-of-way.

Underground utility density is also an important consideration. Refer to ASCE/CI 38, *Standard Guidelines for the Collection and Depiction of Existing Subsurface Utility Data* (2002), for guidance on the collection of existing utility data. The preferred trenchless route and alignment should consider minimizing effects on existing utilities in terms of shaft installation, emergency access, and movement of utilities caused by surface settlement or heave. Trenchless methods in general provide solutions and options not typically available when open-cut trench construction is used.

### 3.1.4 Shafts and Working Space

Pilot tube and other guided boring operations proceed from a jacking shaft to a receiving shaft. Shafts are required to launch the pilot tube and other guided boring methods' equipment, jack the pipe, and retrieve the pilot tube and boring equipment. Shafts in which pilot tubes and the associated pilot tube equipment are used will typically be small, on the order of approximately 9 to 12 ft (2.7 to 3.7 m) in diameter or in a rectangular shape. The jacking shaft will need to be larger when the pilot tubes are used to guide other boring and trenchless methods or when installing longer pipe lengths.

During the planning stage, working space on the surface to accommodate support equipment adjacent to the jacking shafts should be identified. Space is required for:

- Shaft construction, including staging and handling of shoring materials and excavated spoils;
- Power source;
- Lifting equipment;
- Pipe storage and handling;
- Temporary muck handling and storage; and
- Other equipment necessary to support the pilot tube and other guided boring method operations.

Jacking shafts should be of sufficient size to accommodate the earth support system, the jacking system required for the pipe diameter being installed, shaft access devices for personnel, safety equipment, tools, and personnel work space.

Receiving shafts need space for lifting equipment to retrieve pilot tubes, racks, upsizing tooling, and any powered reamer or cutter head equipment.

To the extent possible, shaft locations should be selected that avoid overhead and subsurface utilities. For all shafts, maintenance of traffic during construction (including access for emergency vehicles) should be considered.

Shafts are typically located at changes in alignment (both horizontal and vertical), changes in pipe diameter, and changes in geotechnical conditions and geology. Shaft spacing and locations should also be based on practical drive lengths (see Table 1-1) and project design constraints. The location of shafts may be dictated by the required distance between maintenance access points/manholes to work with the owner's cleaning and inspection equipment.

Shafts constructed far in advance of the pipe installation can be covered if necessary to address public safety or traffic concerns. Excavation of the jacking and receiving shafts in advance of pipeline construction is useful in confirming geotechnical subsurface conditions before jacking pipe. Early excavation of shafts should be considered when complex or difficult ground conditions are expected.

The shaft construction generally affects the public longer than the actual pipeline installation operations and can present risks to the owner, such as settlement, break-in/break-out failures, shaft wall-to-floor failures, and utility disruptions. In addition, when locating shafts, utilities (overhead and underground), settlement of sensitive features, traffic disruptions, vibrations during shaft construction, and disruptions to residences and businesses should be considered.

A jacking shaft is in constant use and contains crews and hydraulic and electrical equipment that must be protected. The jacking shaft should be protected from flooding and cannot normally be covered, other than with a removable security cover during nonworking hours. The dimensions of the jacking shaft need to take into account walers, bracing, thrust blocks, entry/exit seals, ladders, ventilation ducts, sump pumps, jacking frame length, boring equipment, and the pipe segment length. Vertical space above and adjacent to the shaft should be free of overhead utilities. All shafts will need crane access to build the shaft and then position and remove the trenchless equipment, pipe materials, and support equipment. Analyses of safety hazards should be completed for all critical lifts. Overhead wires should be identified before mobilization of equipment. Particular attention should be paid to the latest Occupational Safety and Health Administration (OSHA) regulations regarding maintaining safe distance between cranes and lifting equipment and overhead power lines.

The size of the staging area will vary between shaft construction and pipe installation. During shaft construction, access is required completely around the shaft, the working space is larger, and the working space could require closure of multiple traffic lanes. For planning purposes, the engineer should allow for staging areas of 3,000 sq. ft (280 sq. m) for jacking shafts and 2,000 sq. ft (190 sq. m) for receiving shafts.

During pipe installation, the jacking shaft staging area needs space for cranes, loaders, the power plant, spoils removal, utility pipes, jacking pipe, and trucks. The staging area adjacent to the receiving shaft needs space for cranes, retrieval of pilot tubes and other boring equipment, spoils removal, and trucks. These staging areas will be smaller than that needed for shaft construction. Staging areas may be further reduced in size by storing pipe in other areas and using gantry cranes. Effects of work zones on traffic and other surface activities should be considered.

In planning the geotechnical investigation related to shaft construction, consider that shaft support elements often extend below the actual pipe elevation. Geotechnical borings should extend below the shaft and pipeline depth. When complex geology is present, consider locating shafts at the interface between different geological features, to allow for changes in trenchless construction methods to better suit the changing geology.

The drive lengths should be based on calculations performed by a professional engineer familiar with the process and should consider the project geology, pipe diameter, pipe materials, tracking equipment capabilities and accuracy, cutter tool wear and survivability, and known buried objects. Engineers not familiar with the pilot tube and other guided boring methods should consult with equipment manufacturers and contractors, and review published case histories to establish drive lengths for the pipe diameter being considered.

### 3.1.5 Sustainability

The pilot tube and other guided boring methods minimize environmental impacts (such as noise and dust), minimize the volume of excavated materials, and reduce traffic impacts (such as traffic delays and traffic detours). The pilot tube and other guided boring methods provide greater sustainability benefits for owners and society than do construction methods such as open-trench construction. Therefore, the pilot tube and other guided boring methods should be considered during the planning phase of a civil underground infrastructure project. See ASCE's Policy Statement on Sustainability for more information.

Sustainability comprises a set of environmental, economic, and social conditions in which society as a whole has the capacity and the opportunity to maintain and improve the quality of life indefinitely, without degrading the quantity, quality, or availability of natural resources and ecosystems.

It is ASCE's policy to promote the goal of sustainable civil infrastructure. As stewards of the built environment, civil engineers must lead the way in planning, designing, and constructing sustainable infrastructure. The pilot tube and other guided boring methods offer a sustainable alternative to open-trench construction by avoiding disturbances of ground surfaces, which would require reworking of roads, landscapes, or buildings. When possible, the spoils from the trenchless excavation should be used directly (or indirectly) as backfill.

The carbon footprint of a trenchless project is typically less than that of open-cut construction methods, and can be calculated using a carbon footprint calculator. Carbon footprint calculators for construction are available on the internet. The volume of excavated material is far less with pilot tube and other guided boring methods than with open-trench construction, and this makes trenchless methods a sustainable alternative.

### 3.1.6 Regulations

The engineer should identify and meet all federal, state, and local regulations when planning and designing the project. Procedures to identify and meet regulations should be incorporated into the planning process for a project. At a minimum, the engineer should identify the following:

- Need for environmental documentation and permits;
- Local regulations for discharges and handling of materials;
- Erosion control measures;
- Noise limitations; and
- Work hour limitations.

## 3.2 SITE CONDITIONS

The owner and the engineer should visit the proposed site during the initial planning phase and periodically throughout design to identify site conditions and constraints. A survey of existing site conditions, both surface and subsurface, should be performed in the early stages of planning. This information is useful in determining the feasibility, alignment, grade, and physical constraints of the proposed pipeline project. This work is required independent of the construction method selected. Individuals who understand constructability issues should be consulted and included in the site visits.

### 3.2.1 Land Use

To provide insight into anticipated subsurface conditions, potential future conflicts, and other potential problems, an investigation should be undertaken into the history of the area proposed for the pilot tube and other guided boring operation as well as current zoning and plans for future development. A review of historical drawings, construction records, land-use maps, and aerial photographs can provide insight into past land use that might affect the feasibility and related risk of using pilot tube and other guided boring methods. Review of aerial photographs taken in different eras can be particularly useful in assessing changes in land use and investigating evidence of previous construction. A survey of the project corridor must be done to identify existing infrastructure above and below ground level. Existing property access and traffic conflicts along the corridor need to be identified, reviewed, and understood. To the extent possible, potential future land uses that could restrict the location of structures or affect the alignment need to be identified. Additional site visits may be required in areas of rapid development.

### 3.2.2 Easements and Rights-of-Way

The availability of easements, rights-of-way, and, if necessary, property acquisitions should be determined during the planning process. Early determination of unacceptable corridors may require changes in the alignment. Alignment changes are typically less costly and less disruptive to the project schedule when done early in the planning process. For some pipeline projects, the pilot tube method, other guided boring methods, or another trenchless approach may be the only acceptable method for construction within an easement.

The project owner and the engineer should evaluate different alignments. Consider keeping the alignment alternatives within the public rights-of-way rather than extending them onto private property, which would require acquisition or easement. The cost and schedule of acquisition of a private property or easement must be considered. A comparison of property/ easement cost versus potentially longer and more challenging alignments staying within public rights-of-way should also consider long-term maintenance cost. Efforts to acquire the needed private rights-of-way or easements should start as early as feasible to avoid delaying the project. Acquisition of easements and rights-of-way is often critical to the construction schedule.

Permanent easements for shafts and surface structures should take into account future maintenance and repair needs. Temporary easements are often required for support equipment and storage of materials during construction. A drawing of the site (with all support materials, equipment layout, and shaft locations) should be developed to aid in determining the construction footprint and property needs.

Depth of the pipeline and geotechnical conditions should be considered when evaluating future impacts and property easement costs. When reviewing and acquiring easements, consider widths for future maintenance or repair needs and the most likely inspection and repair methods that will be used. Pipes installed by the pilot tube and other guided boring methods are typically inspected with remotely controlled cameras and repaired with some form of trenchless method. The diameter of the pipe installed by the pilot tube and other guided boring methods often dictates the trenchless repair method, which in turn dictates future needs for access and permanent easement.

In some jurisdictions, a subsurface easement can be obtained rather than acquiring the property or obtaining a surface easement. Obtaining subsurface easements is sometimes easier and less costly than surface easements, but they come with restrictions. When a subsurface easement is obtained, a temporary construction easement may also be required for access to the tunnel horizon from the surface. Subsurface easements could also affect future maintenance and operations.

### 3.2.3 Underground Utilities and Other Facilities

Underground utilities and other facilities along or crossing the pilot tube and other guided boring methods alignment must be identified during the planning process in accordance with state utility-protection laws. Use of ASCE/CI 38 (2002) is recommended, along with the Common Ground Alliance's (2017) *Best Practices Guide.* Utilities and other facilities within the proposed alignment can be affected in a number of ways depending on geology, adjacent underground structures, and material of construction of the existing utility. During the planning phase, it is important to identify those utilities that will have a significant cost, time, or *fatal flaw* impact on the tunneling component of the project. Professional judgment is required to determine just how much utility investigation is necessary, but at the very least, identifying known and suspected utility owners within the pipeline easement and asking them for their records is an important first step. At a minimum per ASCE/CI 38 (2002), a Quality Level D (QLD, records research) utility map should be produced. If an accurate topographic plan has been generated at this stage of the project, this QLD utility map can be upgraded to a Quality Level C (QLC, records research correlated to an accurate site survey) version at minimal cost and time. If high-risk utilities are identified through this process, further actions [such as upgrading specific utilities to Quality Level B (QLB, surface geophysical methods) and Quality Level A (QLA, exposure and test holes)] may be prudent.

Some states have a one-call planning or design ticket. In some cases, using these services may supplement or replace the use of ASCE/CI 38, which may not identify all utilities or provide the detail necessary for the exercise of prudent professional judgment. The engineer must also consider that other objects (such as abandoned utilities, concrete thrust blocks, concrete encasement, manholes, valves, and vaults) may affect the project alignment. Trying to install a pilot tube and other guided bore through previous excavations may have an adverse effect on the grade of the pipeline, even if there is no direct conflict with an existing utility. For example, backfill material or bedding from an existing trench could create a mixed-face condition in an otherwise homogeneous material.

The use of the QLD map, the QLC map, and/or the one-call ticket allows the engineer to make planning decisions on alignment, depth, and necessary utility relocations prior to tunneling, as well as other risk-management issues.

In some cases, a more complete subsurface investigation may be warranted. The combined use of geophysics and test holes may be prudent. ASCE/CI 38 (2002) contains information on the current capabilities of advanced geophysics to image utilities and other features. However, this step of utility identification is usually reserved for the design stage of the project. Particular care should be taken and additional locating effort should be made when crossing any utilities installed by the HDD method because of the variable installation depths inherent with the method.

### 3.2.4 Buried Objects

The potential for encountering buried objects along or crossing the pilot tube and other guided bore should be identified and evaluated during the planning process. The pipeline alignment should be adjusted to avoid known buried objects. Buried objects should be defined, and methods for resolution addressed, in the planning process. Additional assessments of potential buried objects within the planned pilot tube and other guided bore horizon should be undertaken during the geotechnical investigation phase.

Buried objects can be remnants from past construction or natural in origin. Examples of remnants from past construction are: building tiebacks; debris infill; abandoned utilities; wood, concrete, or steel piles; old foundations; utility trenches; and past excavations. Examples of naturally occurring objects are: trees; tree trunks; root systems; cobbles; boulders; voids filled with material that differs, materially and substantially, from the surrounding ground conditions; and flood debris.

A buried object encountered during the work that stops or impedes the installation of the pilot tube and other guided bore, along its design path within permissible tolerances, is an obstruction. If a pilot tube encounters an obstruction, the pilot tube can be retracted from the ground and reinserted. A new alignment is selected, typically moving horizontally to the left or right of the first installation. Usually, the vertical position and grade remain unchanged. A shaft or other means could be used to remove the buried object to allow forward progress of the pilot tube along the original alignment. If larger-diameter boring equipment encounters an obstruction, then a rescue shaft or other means could be required to recover the boring equipment.

### 3.2.5 Contaminated Ground

The type of contamination, location, degree of contamination, and extent of contaminated ground should be determined to assess the effect of the contamination on the final alignment and construction materials, including pipe and joint materials. Contaminated ground should be avoided, if possible. Because the use of the pilot tube and other guided boring methods is typically done above the groundwater table, encountering contaminated groundwater is typically not an issue. Disposal of contaminated ground and groundwater encountered during excavation will increase the overall cost of the project. However, the volume of contaminated material excavated, handled, and disposed of during the trenchless installation is less; it therefore provides an economic advantage over open-cut methods. This advantage could result in an overall lower project cost.

Realignment of the project should be considered to avoid the contaminated ground and to avoid affecting the associated contaminated plume.

Explosive gases and volatile contaminants should be taken into consideration when determining the method of shaft construction, the type of ventilation, and the type of boring equipment that will be used to install the pipeline.

Consultation with specialty environmental and regulatory experts should be considered. Some jurisdictions (national, state, and local) require the preparation of an environmental baseline report (EBR) to document the location of contaminated materials, their potential impact on the environment, and required mitigation methods. The EBR should be considered a contract document, if used.

### 3.2.6 Special Considerations

Special project considerations should be identified early in the planning process. Special considerations may include those listed here and discussed subsequently:

- Limited work hours;
- Restricted access;
- Environmental restrictions;
- Traffic disruptions;
- Haul routes and disposal sites;
- Service interruptions;
- Business disruptions;
- Noise limitations;
- Dust control;
- Light limitations;
- Visual impacts;
- Odor control;
- Potential settlement of existing facilities;
- Utilities underground and overhead;
- Disposal of lubricant, spoils, and groundwater; and
- Public events and construction moratoriums.

*Limited work hours:* Projects located in residential areas may be subject to work hour restrictions by local ordinance. Work hours are typically restricted to nonpeak travel times when work is being done directly in the roadway.

*Restricted access:* The pilot tube and other guided boring methods are focused around the shaft locations. Shafts should be positioned in areas offering access for trucks, employees, and supplies. The shaft sites should be fenced, with the general public restricted from entry. Anyone entering a construction site should be required to check in with the contractor and wear personal protective equipment (PPE) while on site.

*Environmental restrictions:* The pilot tube and other guided boring methods can be located near environmentally sensitive areas. Shafts and associated work sites should be located in such a way as to avoid the defined limits of the environmental area. Some environmental areas may have nesting birds or other wildlife that should be considered and protected. If an environmental document has been prepared for the project, the document may contain mitigation requirements to be followed. Restrictions may apply to the time of year when the work can occur.

*Traffic disruptions:* The pilot tube and other guided boring methods and the associated shafts are typically located within public rights-of way or in city streets. Because the equipment is modular, it can often be set up in a single traffic lane; however, access during construction typically requires daily closing of an adjacent traffic lane. Occupation of the traffic lane requires redirection of traffic past the shafts and the work sites. Management of traffic can disrupt traffic flow. Traffic control plans should be developed for all projects located in city streets. The engineer of the project should look for shaft locations that minimize traffic disruptions or consider traffic detours when appropriate.

*Haul routes and disposal sites:* Spoils from the excavation process will be generated and must be disposed of. Owners may have to identify potential disposal sites for the contractor, especially if contaminated or hazardous material is likely to be encountered. Once a disposal site is identified, haul routes need to be identified. Additional truck traffic will be generated on a daily basis as trucks enter and leave the jacking shaft sites.

*Service interruptions:* The pilot tube and other guided boring methods are used to install new pipelines. Once the main lines are installed, service connections are added. The process of adding service connections to new lines will cause temporary interruption of services.

*Business disruptions:* Businesses may be located along the alignment of the pipe to be installed using pilot tube and other guided boring methods. The businesses near the shafts will be most affected by the work. Shafts should be located where they will not block driveways or sidewalks and prevent customers from reaching the adjacent businesses. The owner of the project should initiate a public outreach program to keep businesses and residents along the project alignment informed of the project's activities and schedule.

*Noise limitations:* Noise will be generated from the construction equipment. This noise will include the hum from generators and other idling equipment as well as backup alarms on trucks.

*Dust control:* Pilot tube and other guided boring methods involve excavation of the ground. Depending on the moisture content of the spoils, the excavated material can become dry and cause dust to be formed.

If the spoils are too wet, mud and fluid spoils can be deposited on the ground surface, where they will eventually dry out and also form dust. Trucks entering and exiting the sites may transport the dust to surrounding areas. Dust control may include sweeping the road surfaces on a regular basis, using water trucks to dampen the dust, placing covers over the trucks as they exit, and washing the truck tires as they leave the site.

*Light limitations:* The pilot tube and other guided boring methods involve the forward advance of pipe into the ground. Sometimes a 24-hour operation is needed to keep the pipe moving forward. In these circumstances, lights will be needed at the shaft locations to allow work to continue through the night. The setup and orientation of the lights should provide sufficient illumination for the workers without disturbing adjacent residents.

*Visual impacts:* Equipment will be staged around the shaft locations and will be visible for the duration of construction.

*Odor control:* Odors may be noticeable to individuals passing by the work sites. Odors may be natural from organic clays, from sewage, contaminated ground, or construction equipment. Mitigation methods to reduce the effects of odor on the public should be considered based on source.

*Potential settlement of existing facilities:* The use of boring equipment involves excavation of the ground to allow forward advance of jacked pipe. This excavation process may result in overexcavation and settlement of the ground. Existing utilities and facilities can be affected by any settlement that occurs.

*Utilities underground and overhead:* The pilot tube and other guided boring methods are used to install pipelines, typically within the road rights-of-way. Other utilities will be in the same rights-of-way. Shafts installed to gain access to the pipeline horizon may encounter these other utilities. Utilities in conflict with the pilot tube and other guided bore will need to be relocated. The pilot tube and other guided boring methods involve the construction of shafts and the lowering of equipment into the shafts. As a result, cranes are needed during the pipeline installation process. The location of overhead wires relative to the cranes' boom and swing needs to comply with OSHA regulations.

*Disposal of spoils and groundwater:* The boring process generates spoils during excavation. This material must be collected and disposed of. The water content of the material will dictate whether the material can be hauled off site in sealed trucks. Construction water, groundwater, and surface runoff may accumulate in the shafts and need to be pumped to the surface. It is sometimes possible to obtain a permit to discharge water collected from the site into the local storm or sanitary sewer system. The volume of material to be disposed of from the excavation process will be significantly less than that generated by open-cut construction.

The volume of spoils will be limited to that produced by the shaft and pipe excavations.

*Public events and construction moratoriums:* Public events and construction moratoriums can occur at any time of the year. Most are well planned and prepared in advance. Typically, a local government will maintain a list of upcoming events and construction moratoriums. If a public event or construction moratorium is planned for the period when the project is under way, the timing of the contract and the construction should be adjusted to minimize effects on the project.

## 3.3 PRELIMINARY GEOTECHNICAL CONSIDERATIONS

In the planning phase of the project, a preliminary geotechnical investigation should be performed to identify the general and any special subsurface conditions. The extent of the investigation required will vary, depending on known local conditions.

In areas of widespread uniform ground formations, it may be possible to use existing data to identify the ground types likely to be encountered. In areas of mixed, varying, or unknown existing conditions, a preliminary investigation should be undertaken to identify the general types of geotechnical conditions. A second function of the preliminary geotechnical investigation is to identify any environmental concerns.

Initial site investigation work, also known as a *desktop study* (because much of the data can be gathered at the engineer's desk through phone calls and Internet searches), is typically conducted during the planning stage of a project. It should include a site visit and a review of all readily available sources of additional information concerning the site. The sources of additional information include geologic and topographic maps, aerial photographs, past geotechnical or soil survey reports, construction case histories, professional papers about the area, historical documents concerning prior site usage, and discussions with local building officials and local contractors. In general, it is prudent to find out as much as possible about the site before planning and implementing further subsurface investigations.

A qualified engineer familiar with the alignment corridor should determine the extent of the preliminary geotechnical investigation that will be required. Preliminary borings may be needed during the planning phase in order to understand alignment alternatives and fill information gaps. The borings should be deep enough to capture conditions that, if present, would affect construction. Recommendations for a complete geotechnical investigation, with borehole spacing and depths, are discussed in greater detail in Chapter 4.

## 3.4 PROJECT LAYOUT

### 3.4.1 Layout for Efficient Pilot Tube and Other Guided Boring Methods

Pilot tube and other guided boring projects should be laid out with attention to the benefits and constraints unique to this construction process. The following recommendations address some of these concerns:

- Minimize the number of pipe ODs. Changing pipe ODs may require different boring equipment and reamer/cutter heads to be mobilized to the construction site. Minimizing the number of different pipe sizes, while still fulfilling the hydraulic/flow design requirements, will reduce the overall equipment requirements for the project.
- Install pipe in more than one direction from a jacking shaft, when possible. To save money and time, pilot tube and other guided boring operations may be advanced both upstream and downstream from the same jacking shaft (see Fig. 2-39). Receiving shafts are smaller and less costly. When practical, layouts of jacking shaft areas should be planned so that areas difficult to access contain only receiving shafts.
- Optimize drive lengths consistent with manhole spacing, pipe size, pipe material, boring equipment used, and geotechnical conditions.
- Evaluate different depths, pipe sizes, and gradients. Increasing the depth, pipe size, and gradient may not have the same effect on a pilot tube and other guided boring methods project as it would on projects using open-cut construction methods. Generally, the cost of constructing a deeper shaft is the only added cost associated with increased depth, provided the pipeline horizon is still above the groundwater table.
- Optimize alignment. Optimum alignment requires the smallest number of shafts without exceeding the practical drive length of the pilot tube and other guided boring equipment and the jacking capacity of the pipe being used, and it follows the most direct route.

Shaft locations will generally be at maintenance access point/manhole locations, at changes in direction, at high and low points of a pressure line, and at other permanent structures located along the pipeline. Local operation and maintenance practices may require additional appurtenances. When local agency or owner constraints limit spacing of manholes and appurtenances, it may be practical to reduce drive lengths to accommodate maintenance or other limitations.

Locations of shafts should be considered with respect to their effects on surface activities. See Chapter 5 for more discussion of shafts and shaft designs. If possible, the route should avoid nondisplaceable soils or rock ground conditions. Consider using a drill bit for rock or nondisplaceable soil conditions as discussed in Section 2.5.1. Mixed-face (soil/rock) conditions should be identified and avoided when using a pilot tube whether with the slanted steering head or rock bit.

### 3.4.2 Alignment and Distance Considerations

Pipe layouts that use pilot tube and other guided boring methods have alignments consisting of straight drives. Slope (grade) will remain constant for the length of the drive. Differences in elevation between individual drives should be accommodated at the shafts.

Choose vertical alignments once end-point requirements have been met. Consider raising or lowering the pipeline profile to avoid unfavorable ground conditions. When evaluating alignment alternatives, consider the predicted settlement trough and settlement-sensitive types of structures that fall within the settlement trough.

The length of a pilot tube and other guided boring installation should be evaluated relative to the risk of completing the drive. Risks to consider include exceeding the jacking capacity of the pipe, losing visual contact with the light source, hitting buried objects, and reducing the effective life of the cutting tools on the equipment.

### 3.4.3 Service Connections

Service connections should be made at manholes and structures to minimize initial cost as well as the cost of future maintenance. Historically, pipelines have been laid in trenches. Service connections (drainage and sewer laterals) were made at any location along the pipeline because access was available during construction. Pipelines with multiple service connections tend to allow the infiltration of groundwater over time, however, and can be difficult to inspect and rehabilitate.

Pilot tube and other guided boring methods allow for the installation of a new pipeline from a jacking shaft to a receiving shaft. Such shafts are the only access points to the new pipeline and are typically converted into manholes. In an effort to reduce the number of service connections between the shafts, the manholes become the primary connection points. Service connections can be made to the manhole in any direction and from any desired depth (see Fig. 2-41). This radiating pattern of service connections to a manhole was first developed in Berlin, Germany, and is often referred to in the literature as the *Berlin construction method,* see Section 2.6.3.

Where connections directly to manholes are not allowed, or when such connections are not feasible, shallow collection lines constructed directly above the new pipeline, or offset and parallel to the new pipeline, can be used to convey local services to manholes. These shallow collection lines are referred to as *riders*. In some instances, other physical constraints, such as numerous utility conflicts, may require connections to be made directly to the new pipeline. If that is the case, excavations are required from the surface to the pipe at the point of connection.

### 3.4.4 Effects on Adjacent Structures

Adjacent structures should be identified, and the effect thereon of pipeline installation in close proximity should be determined. Adjacent structures include aboveground buildings, substructures, vaults, and utilities. The evaluation should include the horizontal and vertical position relative to the pipeline. The zone of influence of both the pipeline installation and working shafts should be identified. The age and condition of adjacent structures and utilities should be considered along with the materials of their construction, their construction history, their sensitivity to ground movement, and their foundation types. The effects of nearby pipeline installation can include ground movements or changes in ground pressure.

With proper planning, pilot tube and other guided boring methods typically have minimal or no impact on structures or utilities adjacent to the alignment. In addition, objects in the structure backfill, or very dense backfill adjacent to the structure, might adversely affect the pipeline installation and/or the structure. Ground movement in soft ground conditions during the boring operation must also be considered. Monitoring of the actual volume of spoils recovered and comparison of recovered material with the theoretical bulked volume may be helpful. The theoretical volume of excavated material should include a bulking factor.

A survey of adjacent structures should be undertaken prior to construction. The survey should include photographs and videos of existing structures, including any visible cracks, settlement, or heave in existing ground surface. When very sensitive or high-risk structures are within a specific zone of influence, the installation and use of geotechnical instrumentation during construction may be required to provide early detection of ground movement so that corrective action can be implemented.

With respect to existing utilities, their material construction and age should be considered. For example, older pipe may collapse or fail if the ground surrounding the pipe moves. Older pipe with unrestrained joints might move, and its joints might open. Welded steel, fused plastic pipe, electrical distribution cables, and communication cables can generally tolerate some movement.

For planning purposes, the zone of influence is initially considered to be a 1:1 horizontal:vertical slope for most geological conditions, but the ratio might be higher or lower. For example, loose granular soil or very soft clays may have a zone of influence of 2:1. For very dense or stiff soils or rock, the zone of influence may be reduced to 0.5:1. The engineer of the trenchless project should consult an experienced geotechnical engineer to determine the zone of influence on the basis of the project's geology.

## 3.5 COST CONSIDERATIONS

Cost is an important consideration when comparing pilot tube and other guided boring methods with other construction methods. Direct costs should not be the only consideration. Pilot tube and other guided boring methods have other benefits that should be considered when evaluating construction options.

Pilot tube and other guided boring methods can be a more economical choice than traditional open-cut construction when all costs are included. The lower direct costs that increase the competitiveness of pilot tube and other guided boring methods are related to the reduced size of the excavated volume, minimizing the amount of surface disruption; simplified traffic control; and reduced project restoration. Other benefits that should be considered include reduced traffic effects and reduced environmental impact.

### 3.5.1 Direct Cost of Pilot Tube and Other Guided Boring Methods

The direct cost of pilot tube and other guided boring methods is influenced or controlled by several factors. These factors include the following:

- Geotechnical conditions and geology and whether they are well documented for the bidders;
- Pipe diameter;
- Objects along the alignment;
- Groundwater;
- Number of competing projects and number of qualified bidders;
- Labor conditions;
- Complexity of the project;
- Location;
- Depth and diameter of shafts;
- Total footage of installed pipeline;
- Mobilization, setup, and moving costs;
- Make-up time;
- Spoils handling and disposal;
- Effect of the pilot tube and other guided bores on existing utilities and structures;
- Special requirements and/or restrictions;
- Project owner's reputation; and
- Availability of suitable equipment for pilot tube and other guided boring methods.

*Geotechnical conditions and geology:* Pilot tube and other guided boring methods may be used in a variety of ground conditions. Different types of ground conditions require different tooling. Costs will potentially increase as the ground becomes harder because it takes longer to excavate the ground. Squeezing ground or gravels/cobbles/boulders increase the probability of becoming stuck in the ground, thus increasing risk and cost. If there is a lack of geotechnical information, bidders may include contingency costs in their proposals that will increase the contract price. The anticipated geotechnical conditions should be clearly stated in the bid documents in order to avoid these contingency costs.

*Pipe diameter:* Large-diameter bores will generally cost more because they require larger and additional equipment, generate more spoils, and require larger-diameter pipe. Because larger boring equipment excavates more ground per foot of advance, such excavations will require larger spoils removal equipment.

*Objects along the alignment:* Hitting one or more objects can slow or stop the forward progress of the equipment and cause delays. Delays mean increased construction time, which in turn increases costs. When buried objects are encountered, additional costs will be incurred for the construction of rescue shafts or for the use of alternative tunneling methods. Even objects that do not stop the equipment of the pilot tube and other guided boring methods can deflect the installation away from its design alignment, damage equipment, and/or accelerate wear.

*Groundwater:* Generally, the pilot tube and other guided boring methods are not used when high groundwater head is encountered because groundwater can flow into the face, carrying soil particles with it, and leading to possible voids in the ground. Should groundwater be present, this may require the use of special equipment, dewatering, more attention to shaft construction, shaft and pipe joint watertightness, and sealing details. Costs of dewatering, pumping, and disposal of groundwater increase with increasing shaft depth. Costs may also increase because of loss of production.

*Number of competing projects and number of qualified bidders:* The number of projects under construction, recently awarded, and in the bidding phase will affect the backlog of the contractors. Contractors who have no backlog or a minimal backlog may set their bid prices lower just to have work for their employees. As contractors' backlogs increase, they can be more selective about which projects to bid and the prices they set for the work. The level of a project's risk, whether real or perceived, will be critical to their interest in bidding. The number of qualified bidders will also affect costs. Increased competition tends to reduce bid prices.

*Labor conditions:* The location of the project, the number of concurrent projects within a geographical area, and whether the area is subject to union rates or prevailing wage rates, will influence the number of workers and their costs.

*Complexity of the project:* As the complexity of the project increases, so will the project's risks and pricing. Increased complexity may reduce the number of bidders and drive costs up.

*Location:* If a project is located close to any established contractors for pilot tube and other guided boring methods, the mobilization costs may be lower. If a pilot tube and other guided boring methods contractor is on a local project, then mobilization costs are lower for other projects in the area if similar equipment can be used.

*Depth and diameter of shafts:* The cost of a shaft incrementally increases with its depth and diameter, presence of groundwater, issues related to ground conditions, and complexity of site access. Shaft shape may also increase cost. Going from a circular to a rectangular shape has the additional cost of the internal support system (walers, struts, and bracing) that will be needed. In larger-diameter shafts, the increased volume of excavated ground for a circular shaft may offset the cost of the internal support systems for a rectangular shaft.

*Total footage of installed pipeline:* As the total footage of a project increases, the unit cost per foot often decreases. The first drive requires mobilization of equipment, construction of two shafts, and setup of all equipment. The second drive may require only one additional receiving shaft and a turn of the jacking equipment in the jacking shaft. The second drive will have a lower unit cost than the first drive, because the costs of the mobilization and jacking shaft are now spread across the two drives. Once the equipment for a pilot tube and other guided boring excavation is on site, and ground conditions remain relatively constant, the incremental unit cost for the excavation tends to decrease because the initial costs are spread over more footage, and the crew tends to become more productive through repetition of tasks.

*Mobilization, setup, and moving costs:* A major cost of a pilot tube and other guided boring methods project is the equipment mobilization, initial setup, and total number of equipment moves. Mobilization costs include preparation of the equipment with new tooling, shop testing and refurbishing of ancillary equipment, and transport of the equipment to the job site. Setting up all of the equipment and spoils handling can take one day or more, depending on the complexity of the project. To move all the equipment to a new shaft location takes time. If two or more drives are possible from the same jacking shaft, the process of preparing and turning around the jacking frame for the additional drives can be done in less time. Consequently, it is most economical to design for drives that are consistent with the equipment's capability, that minimize the number of required equipment setups, and that maximize the number of shafts from which multiple drives can be executed.

*Makeup time:* One component of the advance rate is the makeup time, which is crew dependent. Makeup time is the time it takes to retract the jacks, install a thrust casing or jacking pipe, reconnect the boring equipment, and restart the operations. It also includes the welding time for steel casing. When shorter pipe lengths are used, the number of pipe sections increases. Longer pipe sections require fewer joints to be made up, but also require the construction of longer jacking shafts. So the cost of makeup time must be evaluated against the increased length of the jacking shaft and the project's longer duration. Makeup time is independent of the advance rate.

*Spoils handling and disposal:* Spoils from the bore require handling, including removal from the thrust casing, removal from the shaft, and disposal from the site. Costs will increase with increased bore size and length. All spoils generated at the site require disposal. Increasing the distance to the final disposal site increases cost. Contaminated ground increases the cost of hauling and disposal.

*Effect of the pilot tube and other guided bores on existing utilities and structures:* The condition of the existing utilities or structures needs to be assessed. Utilities may require relocation, replacement, or incorporation of protective measures. Structures may require protective measures.

*Special requirements and/or restrictions:* Any special requirements imposed on a project increase the project's costs. Owners and engineers should work with third parties to minimize such special requirements. Typical special requirements or restrictions may include the following:

- Limits on work spaces;
- Limits on work hours;
- Clearance of road surfaces during peak commute hours;
- Limits on truck trips, pipe delivery, or spoils removal;
- Limits on noise; and
- Dust control and frequency of road cleaning.

*Project owner's reputation:* An owner may have a good or bad reputation regarding contracting requirements and risk sharing. This can increase or decrease a contractor's contingencies on a project, and even its willingness to bid the work.

*Availability of suitable pilot tube and other guided boring methods equipment:* Manufacturers for pilot tube and other guided boring methods typically set the diameter and capabilities of the equipment. Therefore, the size of the casing or jacked pipe becomes critical to the question of which boring equipment is available.

### 3.5.2 Indirect Social Effect Costs

The indirect social effect costs of a construction method should be factored into the process of selecting the construction method. Processes to incorporate these costs have been documented in Boyce and Bried (1994, 1998); Vickridge, Ling, and Read (1992); and Sterling (1994). When alignments are through areas with heavy traffic, the indirect cost to the public may be the overriding consideration in choosing the pilot tube and other guided boring methods. The project affects an area from the shaft construction, through pipe installation, to site restoration. Properly informing the potentially affected parties about the duration and impacts before construction can reduce the costs of mitigating indirect social impacts during construction or prevent social impacts from becoming direct costs.

The cost of traffic disruptions is one indirect cost that can be easily considered. At a minimum, the engineer should use both the planned pipeline route and the proposed detour routes, if applicable, to determine the number of vehicles affected by construction in the road, the potential detour routes, and traffic volumes.

Indirect or social effects to be considered include the following:

- Vehicle traffic disruption;
- Pedestrian disruption;
- Disruption to business activities;
- Loss of tax revenue;
- Disruption to residences;
- Carbon footprint;
- Noise;
- Dust;
- Odors;
- Restoration of pavements; and
- Public safety.

*Vehicle traffic disruption:* When large numbers of vehicles are slowed by pipeline construction in a roadway, the time lost by the drivers and passengers is considered a social effect. There is a social expense in terms of people's wasted time, which increases the anger of motorists and the potential for accidents. During a construction project, an owner or municipality is likely to receive and will need to deal with more complaints from the public. Pilot tube and other guided boring methods minimize the number of disturbed areas in the roadway, thereby minimizing vehicle disruption. One way to accomplish this is to set up work areas with a linear footprint. Pilot tube and other guided boring methods are more effective at reducing vehicle disruption than other construction methods such as open-trench construction.

*Pedestrian disruption:* When large numbers of pedestrians are affected by pipeline construction, the time lost is considered a social effect. There is a social expense in terms of lost personal time and increased potential for accidents. During a construction project, an owner or municipality is likely to receive and will need to deal with more complaints from the public. Pilot tube and other guided boring methods minimize the number of disturbed areas along sidewalks, thereby minimizing pedestrian disruption. If a sidewalk closure is needed, provisions should be made for pedestrians to use crosswalks on both sides of the work space, with plenty of signage to direct them to the designated safe walkway.

*Disruption to business activities:* Depending on the setup of the work spaces, access to businesses may be temporarily disrupted. Pilot tube and other guided boring methods projects are typically positioned within the roadway. Although parking spaces in front of stores may be blocked during construction, access to the businesses and storefronts should remain open. Signage may be needed to let pedestrians and drivers know that businesses have remained open. Shafts should be positioned so as not to block the driveways or entries to parking lots used or needed by the businesses.

*Loss of tax revenue:* Loss of tax revenue, although considered by some to be an indirect social cost, actually is not, and therefore it should not be considered. Consumers buy products when and where they want or need. If access is restricted to a particular store, then the consumer will simply purchase the product at a different location. The amount of tax revenue generated and collected for a government entity is independent of where a product is purchased.

*Disruption to residences:* With any construction, there will be disruptions to the public. Trenchless methods, including the pilot tube and other guided boring methods, are less disruptive than open-cut construction. Any effect related to social costs has to do with the loss of time to pedestrians and vehicles, as discussed above. Complaints concerning the amount of disruption to a specific neighborhood will typically be a function of the influence the residents have with the local government.

*Carbon footprint:* The amount of carbon dioxide generated by a project is now a consideration. As a result, programs exist to determine a project's carbon footprint. Pilot tube and other guided boring methods use fewer surface vehicles, and historically they have generated a lower volume of carbon emissions than open-cut construction.

*Noise:* With any construction, noise will be generated. The issue here is any noise above normal background noise. The noise generated by pilot tube and other guided boring methods typically comes from a single source and can be mitigated with the selection of equipment, and/or operational changes to the equipment. Local ordinances may affect whether work can be completed during off-hours. Noise reduction mitigation measures are a direct cost to the contractor, which is paid by the owner.

*Dust:* Dust is generated because of the excavation of spoils and the movement of equipment. Dust is mitigated by the use of water trucks, covered dump trucks, and tracking mats. Any fugitive dust can be cleaned with street sweepers. Dust reduction measures are a direct cost to the contractor, which is paid by the owner.

*Odors:* Odors can be generated from equipment (fumes). In general, these odors are limited. In some cases, excavations for shafts in contaminated ground may generate odors that can be offensive to the public. In areas in which submerged marine clays may be encountered, disturbance of the marine clay may release odors from decayed marine life. These odors may be difficult to mitigate during construction but can be avoided with proper planning of the alignment.

*Restoration of pavements:* Municipalities as well as regional and state agencies regularly renew pavements. These restoration projects generally consist of full-depth reconstruction, at a significant cost to these agencies. To maintain the structural integrity of the pavement as a whole, these agencies may require full sectional reconstruction of the pavement rather than simply patching the immediate area of disturbance. There is a direct cost to the contractor and, ultimately, to the project sponsor when full sectional reconstruction is required. There is also lost cost to the agency that expended monies to replace the pavement in the first place. Coordination, when possible, with the agency responsible for pavement maintenance should be conducted to adjust schedules and cost. In other words, project work should be done before any planned pavement replacement. In cases in which older pavement is concerned, incisions caused by shaft excavations and by relocation of associated utilities may reduce the structural integrity and life expectancy of the pavement. This disruption may require expenditures of monies sooner than expected by the agency responsible for pavement maintenance.

*Public safety:* Pilot tube and other guided boring methods are some of the least invasive trenchless installation methods because the footprint of the operation can be a relatively small. However, surface support equipment for the construction will have a significant footprint at the jacking shafts. The footprint may affect pedestrian use of sidewalks, reduce travel lane widths, restrict emergency vehicle access, or block access to hydrants or buildings in ways that affect public safety. Coordination with local emergency services should be conducted during the design and construction of the project. There is an associated cost for these additional services, which is generally borne by the local taxpayers.

### 3.5.3 Environmental Costs

The effect of construction through contaminated ground should be taken into account when potentially contaminated ground has been identified during the investigation process. A contingency plan identifying methods for handling contamination may be required as part of the contract documents when construction will be through potentially contaminated areas. When possible, shafts should be located outside the contaminated area or constructed using methods designed to minimize the migration of contaminated ground. Using pilot tube and other guided boring methods below the level of the contaminated material can significantly reduce or eliminate the problems associated with installing a pipeline through these areas.

### 3.5.4 Risk Analysis

A risk register should be started for the project during the planning phase to help reduce overall project costs. The register is a living document and should be updated throughout the project's life. The engineer, with input from the owner, should identify and analyze the risks associated with viable construction methods. It may be helpful in some cases to conduct an alternative analysis of pilot tube, guided boring, and other trenchless methods that accounts for both initial and long-term direct costs, indirect costs, and the impact of construction on third parties.

The process for developing a risk register includes the following activities:

- Hold a brainstorming session, with key members of the design team and the owner, to identify project hazard and risk scenarios for each construction method proposed.
- Group the hazard and risk scenarios into related categories (e.g., shafts, trenchless methods, public outreach, permitting, property acquisition, safety, contractual issues).
- Identify the causes of each identified hazard or risk scenario.
- Determine the probability or likelihood that a specific hazard or risk scenario will occur.
- Determine the potential effect of the occurrence in terms of relative cost and/or scheduling delay to the project.
- Rate the hazard and risk scenarios on the basis of their likelihood and severity.
- Develop mitigation measures to reduce the probability and/or severity of hazard or risk scenarios.
- Rerate the hazard or risk scenarios with mitigation measures incorporated.
- Update the risk register with new identified hazard or risk scenarios as the project advances in design and into construction.

As an example, the severity of the effect and the probability of a risk are both ranked from 1 to 5. Each risk ends up with a ranking between 1 and 25 (see Table 3-1). Risks ranked in the red zone (12 to 25) should be mitigated during design into the yellow zone (5 to 10) or green zone (1 to 4). Risks ranked in the yellow zone should be mitigated during design into the green zone, to the extent possible.

The risk register becomes a project management tool for making sure that mitigation measures, once identified, have been incorporated into the design and construction.

Table 3-1. Risk Table with Severity of Impact and Probability of Occurrence

| Impact \ Probability | Rare (1) | Unlikely (2) | Possible (3) | Likely (4) | Most likely (5) |
|---|---|---|---|---|---|
| Very high (5) | 5 | 10 | 15 | 20 | 25 |
| High (4) | 4 | 8 | 12 | 16 | 20 |
| Moderate (3) | 3 | 6 | 9 | 12 | 15 |
| Low (2) | 2 | 4 | 6 | 8 | 10 |
| Very low (1) | 1 | 2 | 3 | 4 | 5 |

Risks may exist for one or more drives on a trenchless project that could prevent the drive from being completed successfully. The effect of failing to complete a drive should be considered. A contingency plan should be prepared, identifying at least one alternative construction method.

### 3.5.5 Contingency Costs

There are unknown or unforeseen conditions associated with any underground project. Contingency costs should be identified and included in any construction cost budgets prepared by the owner. Despite completion of a risk register, some risks cannot be eliminated or reduced by design or construction mitigations alone. To accommodate the costs associated with these risks, the owner should consider adding contingencies to the construction cost budget. These contingency costs can be held outside the construction contract or included as allowances within the bid schedule. The allowances are fixed dollar amounts, set by the owner, to cover additional items for the project that could not be quantified at the time of the bid. As an example, allowances can be included for disposal of contaminated ground.

The cost of a rescue shaft is also an appropriate allowance item. The design team for the project may know of, or anticipate, potential buried objects within the pipeline alignment that could slow or stop the forward progress of the trenchless installation. However, the exact location, surface accessibility, frequency of use, size, and depth of the rescue shaft are not known. To ensure all bidders include this item, an allowance amount should be included in the bid schedule. When the trenchless equipment becomes stuck, the owner and the contractor can develop a rescue plan. The cost for the rescue shaft can then be established on a time-and-materials basis, with funding provided from the allowance item.

When owners show their willingness to identify and share project risks, the bidding contractors are more likely to include fewer contingencies in their bids.

## REFERENCES

ASCE. (1998). "Pipeline route selection for rural and cross-country pipelines." Manuals and Rep. on Engineering Practice No. 46, Reston, VA (Out of print).

ASCE. (2002). "Standard guidelines for the collection and depiction of existing subsurface utility data." ASCE 38-02, Reston, VA, 32.

Boyce, G. M., and Bried, E. M. (1994). "Benefit-cost analysis of microtunneling in an urban area." *Proc., North American No-Dig 1994*, NASTT, Dallas, TX.

Boyce, G. M., and Bried, E. M. (1998). "Social cost accounting for trenchless projects." *Proc., North American No-Dig 1998*, NASTT, Albuquerque, NM, 3–12.

Common Ground Alliance. (2017). "Best practices guide." ⟨http://commongroundalliance.com/best-practices-guide⟩ Accessed on March 2017.

NASTT (North American Society for Trenchless Technology). (2012). "Development of a carbon calculator tool to compare trenchless and standard pipeline activities." *Proc., No-Dig Show*, Nashville, TN.

Sterling, R. L. (1994). "Indirect costs of utility placement and repair beneath streets." Rep. No. MN/RC-94/20, Minnesota Dept. of Transportation, Minnesota, MN.

Vickridge, I., Ling, D. J., and Read, G. F. (1992). "Evaluating the social costs and setting the charges for road space occupation." *Proc., Int. No-Dig*, ISTT, Washington, DC.

# CHAPTER 4
# SITE INVESTIGATION

## 4.1 GENERAL

This chapter addresses site investigations that are important for the planning, detailed design, construction, long-term operability, and resiliency of pilot tube and other guided boring projects. These site investigations include the following:

- Geotechnical assessment;
- Utility surveys (both aerial and subsurface);
- Traffic flow and access for vehicles and pedestrians;
- Environmental conditions;
- Flood zones; and
- Seismic conditions.

This chapter also includes a table for determining which variations of pilot tube and other guided boring methods are appropriate based on identified ground conditions.

It is not the intent of this section to provide detailed guidance on conducting all site investigation programs or laboratory and field testing for specific pilot tube and other guided boring projects. All projects are different and require site-specific investigations.

## 4.2 GEOTECHNICAL ASSESSMENT

It is important that ground and groundwater conditions and their range of variability be reliably established as early as possible during planning to determine the feasibility of implementing pilot tube and other guided boring methods. If it is determined to be feasible, the information obtained will be used to design and construct the project. Problematic ground conditions can reduce feasibility of, or eliminate, pilot tube and other guided boring methods as an option for construction.

### 4.2.1 Site Investigation

A common and accepted practice is to conduct a detailed site investigation in phases. Each subsequent phase of the study addresses gaps identified in the prior phase of the investigation and provides the information needed to plan the next phase of study.

The first phase is a desktop study, whereas the next phase may consist of a few test borings, supplemented with one or more test pits and/or observation wells with samples obtained for laboratory testing purposes. For a smaller project, this may be the primary design-level investigation. Upon completion of this work, it is then possible to establish a working model for the site including ground properties, stratigraphy, and groundwater conditions. Use of the observation wells will allow the collection of groundwater information to establish the fluctuation of groundwater levels. Groundwater samples obtained from the wells can also be analyzed for water quality and corrosivity if the wells are installed in accordance with current environmental procedures.

A subsequent phase of investigation represents an opportunity to test the validity of the subsurface model. Depending on the outcome of this work and the importance of specific conclusions relative to subsurface conditions, it could be prudent to conduct an additional third phase of even more detailed investigations to resolve remaining ground condition uncertainties. For additional discussion on conducting geotechnical investigations refer to Federal Highway Administration (FHWA 2001). For additional discussion of the phased approach for underground projects, refer to Parker (2004).

**4.2.1.1 Desktop Study.** The desktop study typically consists of a review of existing information, including

- Geologic maps;
- Local or regional groundwater (and surface water) conditions and historic highs and lows;
- Current and historical aerial photography;
- Historical maps and photographs;
- Existing or historical subsurface information, which may include logs of subsurface investigations;
- Construction records from prior nearby projects or projects in similar subsurface conditions; and
- Potential for contamination, including review of state and private website tracking tools and record searches as well as aerial photography, mapping, and on-site photography.

Depending on the scope of the project, the results of the desktop study may be published as a preliminary geotechnical report or memorandum, or included in a preliminary design report. For some projects, the results of the desktop study may simply be used to plan the primary-level subsurface investigation and may be included with the primary geotechnical reporting.

**4.2.1.2 Subsurface Investigation Methods.** Initial site investigation work, which is typically conducted during the planning or preliminary design stage of a project, should include a site visit and a review of material gathered during the desktop study.

The most common form of subsurface investigation is the sampled test boring. This involves advancing a small-diameter vertical borehole using a casing or hollow-stemmed auger, which collects samples of the material encountered. An important first decision is to determine how many test borings to drill and to what depth. Many guidelines include suggestions on number of test borings per foot of pipeline. As a minimum, test borings should be taken at all jacking and receiving shaft locations unless suitable historical data exist. Depending on the subsurface conditions present, test borings at the shaft locations may be sufficient for shorter, less than 200 ft (60 m), pilot tube and other guided boring installations. Additional borings are often appropriate for pilot tube and other guided boring drives longer than 200 ft (60 m).

Test borings at shafts should extend sufficiently below the shaft or pit bottom to identify changes in the strata below both the shaft and the pilot tube and other guided bore alignment that could affect the construction. The additional boring depth allows collection of geotechnical information for the shaft foundation or plug, if applicable. Exploration should extend at least 10 ft (3 m) below the tip elevation of any proposed shaft ground support systems. Test borings along the pilot tube and guided bore alignment should extend to at least 10 ft (3 m) or two pipe diameters below proposed pipe invert, whichever is deeper. If soft, compressible soils are present at or near the invert, some of the borings should be extended into firm ground.

The final decision on the number and depth of borings should be based on the importance and complexity of the project, the anticipated risks, site access, and the nature and complexity of the ground conditions. The scope and results of the subsurface investigation program should be reviewed by geotechnical personnel experienced with the local geologic conditions. The more difficult and costly the proposed installation, the more potential effects to third parties; and the more complex the ground conditions, the more extensive the subsurface investigation program should be.

During completion of the test borings, soil samples are commonly obtained using split spoon samplers, at the end of the standard penetration test (SPT), in accordance with ASTM D1586. This involves driving a hollow steel sampler 18 to 30 in. (450 to 760 mm) long and 1-3/8 in. (35 mm) in diameter at 6-in. (150 mm) intervals, using a 140-pound hammer, falling freely over a vertical distance of 30 in. (760 mm). The procedure involves driving the sampler either 18 or 24 in. (450 or 600 mm) and recording hammer blows in 6-in. (150 mm) increments. The sum of the blows recorded over the 6- to 12-in. (150- to 300-mm) and 12- to 18-in. (300- to 450-mm) intervals is defined as the N-value, and provides a measure of the relative density or consistency of the soil. The reader should refer to ASTM D1586 for further information. For samplers and hammers that deviate from ASTM D1586 (for example ASTM D3550, which was withdrawn in 2016), the reader is cautioned that these field blow counts do not represent N-values. For additional information consult a geotechnical engineer.

Split spoon samples typically should be taken at intervals no greater than 5 ft (1.5 m) and at changes in geologic strata. Continuous sampling within the tunnel horizon is prudent to identify thin strata that could have significant effects on the pilot tube and other guided bore advance. If classifications of soil and rock are the only tests required, then disturbed sampling procedures are acceptable (i.e., split spoon). If, however, laboratory strength and compressibility tests are needed, undisturbed samples may be substituted for split spoons.

If bedrock is encountered, continuous coring should be completed using triple-tube (e.g., HQ, NQ) core barrels. The length of the core run will depend on the rock mass quality, but, in most cases, should not exceed 5 ft (1.5 m). Coring may also be a viable sampling method if oversized materials (e.g., cobbles, boulders) are encountered.

Field tests that might be used include in situ vane testing, hand penetrometer testing, point load testing on rock, pressure meter testing, and in situ downhole permeability or pumping tests.

At some sites, it may be prudent to supplement the test boring program with other forms of subsurface investigation such as test pits, cone penetration tests (CPTs), rotosonic borings, or large-diameter drill holes. Test pits, rotosonic borings, and large-diameter drill holes can be useful methods of subsurface investigation, especially for identifying the presence of gravels, cobbles, and boulders; where subsurface conditions are extremely variable; or in urban settings in which areas of undocumented fill may exist. They may also provide direct evidence of the type and distribution of objects that exist in the ground and reveal important facts about possible ground behavior characteristics such as sloughing, squeezing, and groundwater inflows. CPTs involve pushing a piezocone probe into the ground and can be a cost-effective method of obtaining a continuous record of subsurface conditions. CPTs should always be calibrated at the site with one or more coupled soil borings, and their use is contingent on soil density/hardness and CPT rig capacity.

There are also many innovative and cost-effective methods of investigating the ground by geophysical techniques and other emerging methods. These investigations can be effective in helping to interpolate subsurface conditions between the test borings. Although a complete listing of all possible geophysical exploration methods is beyond the scope of this section, some examples are

- Seismic refraction and reflection surveys,
- Ground-penetrating radar, and
- Electrical resistivity studies.

These methods and downhole investigations have been shown to be effective in some types of subsurface profiles. In general, such applications should be used if they provide useful information at a nominal cost relative to the proposed design and construction requirements. The effectiveness of these methods may depend on the subsurface conditions present. An experienced geophysical professional should be consulted during planning of the geophysical program, to ensure the appropriate methods are employed.

Other issues of importance to the success of a subsurface investigation program include good-quality exploration and sampling services and continuous on-site monitoring of the work by a trained professional representing the engineer. Thorough observation and documentation of field activities must be provided so the results of the field work can be accurately and correctly interpreted.

### 4.2.2 Characterization

The following section describes site characteristics that are important to identify and quantify during the site investigation as they may affect the effectiveness of the pilot tube and other guided bore installation. Site characteristics that should be identified for pilot tube and other guided boring projects are

- Environmentally sensitive surface features;
- Surface and subsurface features resulting from past construction;
- Consistency/density of the ground (very soft to hard/very loose to very dense);
- Nature of bedrock (degree of fracturing, weathering, hardness, abrasivity);
- Potential for mixed-face condition;
- Potential for cemented soils/caliche;
- Shallow groundwater conditions;
- Potential buried objects, including utilities;
- Potential for naturally occurring gases such as methane and hydrogen sulfide; and
- Contaminated groundwater or soil.

**4.2.2.1 Soft Ground Conditions.** An evaluation of whether the soil will be readily displaceable is a critical component in assessment of the feasibility of the pilot tube method for an installation. Typically, the density or consistency of the soil can determine whether the ground is considered displaceable or nondisplaceable. A detailed summary of pilot tube and other guided boring applicability in various ground conditions is provided in Section 4.2.3.

For cohesionless soils (predominantly sands and gravels), important properties that should be identified include

- Grain size distribution;
- Unit weight;
- Hydraulic conductivity;
- Density (typically in terms of SPT N-values); and
- Maximum clast size (i.e., gravel, cobbles, and boulders if present).

For cohesive soils (clays and silts), important properties that should be identified include

- Moisture content;
- Atterberg limits (plasticity index and liquid limit);
- Unit weight;
- Shear strength;
- Compressibility;
- Consistency [typically in terms of unconfined compressive strength (UCS) or SPT N-values];
- Grain size distribution sieve and hydrometer analysis; and
- Maximum clast size (i.e., gravel, cobbles, and boulders if present).

Laboratory testing should be done in accordance with the appropriate ASTM standards.

The Tunnelman's ground classification system, first described by Terzaghi (1950) and refined by Heuer (1974), is designed to describe ground behaviors and their effect on larger, conventionally constructed soft ground tunnels. The system is also a useful tool for evaluation of soft ground conditions as they relate to pilot tube and other guided boring methods. Descriptions of various ground behaviors, in terms of the Tunnelman's ground classification system and its effects on pilot tube and other guided boring methods, are presented in Table 4-1. General ground behaviors include raveling, running, flowing, squeezing, and firm ground. See Section 4.2.4 on applicability for further discussion of the importance of the ground classification.

**4.2.2.2 Rock Conditions.** The use of a pilot tube–guided method is feasible in rock with the proper tooling. Rock is a nondisplaceable material so a drill bit is needed to create the initial pilot tube borehole. Consult Section 2.5.1 for additional discussion on rock tooling.

Table 4-1. Tunnelman's Ground Classification for Soils (after Heuer 1974)

| Classification | | Behavior | Typical Soil Types |
|---|---|---|---|
| Firm | | Heading can be advanced without initial support and final lining can be constructed before ground starts to move. | Loess above water table; hard clay, marl, cemented sand, and gravel when not highly overstressed. |
| Raveling | Slow Raveling | Chunks or flakes of material begin to drop out of the arch or walls sometime after the ground has been exposed. This is due to loosening or to overstress and *brittle* fracture (ground separates or breaks along distinct surfaces, as opposed to squeezing ground). | Residual soils or sand with small amounts of binder may be fast raveling below the water table; slow raveling above. Stiff fissured clays may be slow or fast raveling, depending upon degree of overstress. |
| | Fast Raveling | In fast raveling ground, the process starts within a few minutes, otherwise the ground is slow raveling. | |
| Squeezing | | Ground squeezes or extrudes plastically into tunnel, without visible fracturing or loss of continuity, and without perceptible increase in water content. Ductile, plastic yield and flow due to overstress. | Ground with low frictional strength. Rate of squeeze depends on degree of overstress. Occurs at shallow to medium depth in clay of very soft to medium consistency. Stiff to hard clay under high cover may move in combination with raveling at excavation surface and squeezing at depth behind surface. |

*(Continued)*

Table 4-1. Tunnelman's Ground Classification for Soils (after Heuer 1974) *(Continued)*

| Classification | | Behavior | Typical Soil Types |
|---|---|---|---|
| Running | Cohesive Running<br>Running | Granular materials without cohesion are unstable at a slope greater than their angle of repose (±30 to 35°). When exposed at steeper slopes, they run like granulated sugar or dune sand until the slope flattens to the angle of repose. | Clean, dry granular materials. Apparent cohesion in moist sand or weak cementation in any granular soil may allow the material to stand for a brief period of raveling before it breaks down and runs. Such behavior is cohesive running. |
| Flowing | | A mixture of soil and water flows into the tunnel like a viscous fluid. The material can enter the tunnel from the invert as well as from the face, crown, and walls, and can flow for great distances, completely filling the tunnel in some cases. | Below the water table in silt, sand, or gravel without enough clay content to give significant cohesion and plasticity. May also occur in highly sensitive clay when such material is disturbed. |
| Swelling | | Ground absorbs water, increases in volume, and expands slowly into the tunnel. | Highly preconsolidated clay with a plasticity index in excess of about 30, generally containing significant percentages of montmorillonite clay. |

The boundary between soil and bedrock UCS is commonly described as 140 psi (20,000 psf) (Kulhawy et al., 1991) and obviously should be considered nondisplaceable ground. Based on appropriate descriptive criteria [i.e., ASTM, International Society for Rock Mechanics (ISRM)], the field subsurface and surface site investigation strategy for projects in which bedrock is anticipated should focus on determining

- Depth and extent of bedrock;
- Rock type (e.g., lithology, classification);
- Mineralogy;
- Core recovery;
- Rock quality designation (RQD);
- Weathering/alteration index;
- Fracturing frequency;
- Discontinuity set frequency/spacing;
- Discontinuity surface characteristics (e.g., aperture, infill material, roughness, shape, joint roughness coefficient);
- Presence of fracture-controlled groundwater;
- Discontinuity orientation (e.g., dip and dip direction);
- Hardness; and
- Field-estimated UCS (e.g., point load testing, Schmidt rebound).

Conventional rock coring potentially accompanied by downhole acoustic/optical televiewer logging can be a cost-effective way to arrive at the field data outlined above, and also allows the user to collect data on discontinuity orientation.

After the field investigation phase, laboratory testing should be completed on select samples obtained in the field, to arrive at reasonable estimates of

- Hardness (e.g., Shore);
- UCS;
- Modulus of elasticity;
- Poisson's ratio;
- Tensile strength (e.g., Brazilian splitting);
- Punch penetration;
- Intact rock or joint direct shear strength;
- Mineralogy (e.g., thin sections);
- Rock mass rating system (e.g., RMR, Q, GSI);
- Abrasiveness (e.g., Cerchar, Taber, SAT);
- Cutability;
- Boreability; and
- For claystone and shale, slake durability, jar slake, and swelling tendencies.

Drilling and sampling procedures that both maximize core recovery and minimize core damage are essential to the development of a reliable rock mass characterization program. Triple-tube (e.g., HQ, NQ) coring generally provides the least sample disturbance.

When collecting rock cores, it is wise to take photographs with a scale and color guide immediately upon opening the sample. Once the sample is opened or moved it will begin to change from its in situ condition because of exposure to the atmosphere and vibration. It is also vital that photographs of each completed core box be obtained. The photos must show core intervals clearly.

Point load tests and Schmidt hammer tests can be used to supplement laboratory-determined UCS data and to rapidly obtain preliminary strength data in the field.

Weathering index, fracture frequency, joint characteristics, and overall rock mass quality are also important for estimating boreability. Estimation of block sizes can be an important factor for evaluating the potential for blocks to become wedged in the reaming head or to become separated or dislodged above the pipe and cause gouging of the pipe or couplings.

RQD is widely used as an indicator of overall rock quality. Rock quality determinations can be refined, if necessary, using various empirical rock mass classification schemes as proposed above (e.g., RMR and Q), which were developed for construction of larger tunnels (Bieniawski 1974; Barton et al. 1974).

Boreability is the ability of the rock in front of the pilot tube and cutter head to be ripped, chipped, and broken into smaller pieces to allow the advance of the pipe string. This parameter is controlled by the frequency of discontinuities of the rock; best determined based on RQD (ASTM D6032), and the following four tests performed on intact samples of rock: UCS (ASTM D7012), tensile strength (ASTM D3967), hardness (ASTM D5873), and abrasiveness. The four properties of the intact rock are determined by conducting a series of laboratory tests. The boreability of the rock also affects the tooling efficiency and the tooling survivability. Some rock may be too strong, unfractured, and abrasive to be cutable by typical tooling. Drive lengths may require modification. In some cases, other construction methods should be considered.

**4.2.2.3 Mixed-Face Conditions.** Mixed-face conditions are defined by distinct variations in ground conditions within the cross-sectional area of the bore, such as rock overlain by or interbedded with soft ground or very soft soil overlain by, underlain by, or interbedded with a very stiff, high-strength soil. Mixed-face conditions present significant challenges to alignment, grade control, and stability of the face, over-excavation, and stoppage of the machine advance. Mixed-face conditions should be identified during the design phase to the extent possible and should be avoided if possible.

When advancing from a full face of soft ground into a zone with a mixed face of hard and soft ground, better line-and-grade control can in some cases be achieved by advancing the pilot tube from hard ground into the softer ground.

**4.2.2.4 Gravels, Cobbles, and Boulders.** The extent, frequency, size distribution, maximum clast size, and physical properties of gravels, cobbles, and boulders should be determined. The pilot tube and other guided boring drives should be located to minimize the amount of gravels to be encountered, if possible. If these conditions cannot be avoided, other trenchless methods should be considered. The presence of cobbles and boulders can present a significant challenge to pilot tube and other guided boring methods and may result in a failure to complete the bore.

If rock pieces are encountered within the sampler during the geotechnical investigations, steps should be taken to determine if the rock pieces are naturally occurring gravel or were generated by the sampler encountering and fracturing cobbles and/or boulders. Where there is a potential for gravels, cobbles, and boulders, large-diameter boreholes, rotosonic borings, test pits, or trenches can be useful in obtaining representative samples for grain size analysis and determining the size and distribution of the gravels, cobbles, and boulders. Samples of the cobbles and boulders as well as the matrix material should also be obtained and tested.

One way to document the frequency of the cobbles or boulders is to determine the cobble volume ratio (CVR) and the boulder volume ratio (BVR). The CVR and BVR are the volume of cobbles or boulders to the excavation volume reported in terms of percentage. (See Hunt et al., 2013.)

**4.2.2.5 Groundwater.** The groundwater level in unconfined aquifers and piezometric levels in confined aquifers should be determined by installing observation wells and/or piezometers adjacent to shaft locations. Consider adding piezometers at intermediate points for longer drives. In addition, the hydraulic conductivity of water-bearing strata should also be determined.

Groundwater conditions will have a significant influence on ground behavior (see Table 4-1), the viability of the pilot tube method, and on jacking and receiving shaft design and construction. Hydraulic conductivity of water-bearing strata can be estimated using grain size correlations and preferably by borehole permeability tests. For larger projects, in which highly permeable soils are anticipated with significant groundwater control issues during construction, pumping tests may be warranted.

Dewatering in the immediate vicinity of active pilot tube and other guided boring operations can be an effective way to improve the ground and facilitate construction. It should be noted, however, that dewatering may also have unintended consequences such as increased friction on the pipe, resulting in higher jacking forces because of increases in effective stress and loss of lubrication.

**4.2.2.6 Potential Obstructions.** The likelihood of buried objects, their nature, and relative sizes should be established by the desktop study and site investigation. The use of pilot tube and the ability to withdraw the pilot and attempt a revised alignment is one of the many benefits of the pilot tube method. However, to avoid multiple attempts, the potential for buried objects and an understanding of their nature should be determined early in the planning phase to minimize the risk of encountering an obstruction.

**4.2.2.7 Contaminated or Hazardous Ground or Groundwater.** Encountering ground and/or groundwater contamination has health and safety, cost, and schedule impacts on projects. Hazardous conditions can include naturally occurring hydrocarbons and asbestos, but careful planning and execution may minimize these impacts. Determination of the potential for encountering contaminants and hazardous substances should be completed during the planning phase. If contaminants are found, then determination of the nature and extent of contaminants, if present, must be undertaken during the site investigation. Even if contaminants are not identified during the planning phase, the site investigations should screen for contaminants. One useful approach is to sample the headspace above the samples with a photo-ionizing detector during sampling and to record the readings. Consult an environmental professional for additional information regarding detecting the potential presence of containments in the ground.

### 4.2.3 Geotechnical Reports

All subsurface data collected during the geotechnical desktop study and site investigations, professional interpretations thereof, and design and construction considerations should be summarized in project reports. The geotechnical data report (GDR) contains all the factual geotechnical information for the project, including explorations, laboratory and field testing results, and geophysics and historical geotechnical data. The GDR typically does not include interpretation or recommendations, and should be included in the contract documents.

Geotechnical design memoranda and/or a geotechnical interpretive report (GIR) should be prepared to present summaries of the geotechnical data, interpretation of the data, earth pressures to be used for design, discussions of the expected behavior of the ground, and other geotechnical design recommendations such as appropriate tunneling and shaft types and systems. Because the GIR and design memoranda are typically prepared prior to design, there is much said in these memoranda and reports with respect to the project that may not be applicable at the time of bid. These design memoranda and GIR should be disclosed to the bidders but are typically not provided to the bidders, nor should they be in the contract documents.

Increasingly, another standalone report is prepared, known as a geotechnical baseline report (GBR). The GBR is typically prepared as the design is being completed with the owner's input and serves as the definitive geotechnical baseline for use in the resolution of disagreements, disputes, or claims relating to differing subsurface conditions. The GBR presents contractual interpretations of the data to be used for bidding and construction, as well as baseline expected behavior of the ground, and other geotechnical construction considerations, such as appropriate tunneling and shaft types and systems. See ASCE's *Geotechnical Baseline Reports for Construction: Suggested Guidelines* (ASCE 2007). The GBR should be included in the contract documents and should be prepared by a qualified geotechnical engineer experienced in underground construction.

### 4.2.4 Applicability of Methods Based on Subsurface Conditions and Classification

As discussed previously, the geotechnical conditions can often be grouped into displaceable and nondisplaceable categories. Rock most often classifies as nondisplaceable for obvious reasons. The information in Table 4-2 can be used to identify whether a displaceable or nondisplaceable method is more appropriate for the identified geotechnical conditions, and whether pilot tube in general is a feasible method for the project.

Because pilot tube and other guided boring methods are not considered to be a closed-face trenchless method, they are not considered applicable for use in ground conditions that classify as flowing (per terminology used in Table 4-1). Ground improvement, such as dewatering, can be implemented to modify the ground classification. Similarly, squeezing ground can result in high jacking forces, which may exceed the capacity of the chosen equipment. Identification and mitigation of the risks associated with these ground types can be key to successful completion of a drive.

## 4.3 UTILITY SURVEYS

It is important that the locations of existing utilities be reliably established as early as possible during planning to determine the feasibility of implementing pilot tube and other guided boring methods. Existing utilities should be dealt with in an iterative manner in which each level of survey builds on the prior study and identifies data gaps to be filled in a subsequent study.

Table 4-2. Applicability of Pilot Tube and Other Guided Boring Methods for Identified Geotechnical Conditions

| SPT N-value | Geotechnical condition | Applicability (by method): Displacement | Applicability (by method): Nondisplacement |
|---|---|---|---|
| 0–1[a] | Very soft clays, silt, and organic deposits | Marginal[b] | Marginal[b] |
| 2–4 | Soft clays, silt, and organic deposits or very loose sands | Yes | Yes |
| 5–10 | Medium stiff clays and silts or loose sands | Yes | Yes |
| 11–30 | Stiff to very stiff clays and silts or medium dense sands | Yes | Yes |
| 11–30 | Soil with gravels (1 to 3 in.) | Yes[c] | Yes[c] |
| 11–30 | Soils with occasional cobbles (3 to 12 in.), boulders (>12 in.) | Marginal[d] | Marginal[d] |
| 11–30 | Soils with significant cobbles, boulders, and obstructions larger than 4 in. | No | No |
| 31–50 | Hard clays or dense sands | Marginal[e] | Possible[f] |
| >50 | Very dense sands or weak weathered rock | No | Possible[f] |
| >50 | Weathered rocks, marls, chalks, and firmly cemented soils | No | Possible[f] |
| >50 | Significantly weathered to unweathered rocks | No | Possible[f] |

Note: Generally, larger equipment and tooling (i.e., larger-diameter augers) have a better success rate in more complicated geotechnical conditions.

[a]Also referred to as *weight of hammer material*.

[b]The ground conditions are too weak to maintain line and grade.

[c]As the gravel content increases, the success rate decreases.

[d]Total chance as to whether the drive is completed depends on hitting the buried object.

[e]Pilot tube may be able to penetrate a portion of the ground with this N-value but not for an extended distance.

[f]With the proper rock tooling on the pilot tube and reamer/cutter head.

All utility information (aerial and subsurface) should be collected in connection with the project alignment(s) and presented in accordance with ASCE/CI 38, *Standard Guidelines for the Collection and Depiction of Existing Subsurface Utility Data* (2002). It is recommended to obtain Quality Level B and Quality Level A utility data, especially if the utility expectancy is complex and/or risky. Note that using a *design ticket* from a state's on-call service can only result in Quality Level C or Quality Level D information.

During the planning and design phase, it may be possible to make changes to the bored alignment and shaft locations to avoid conflict and the necessity for a potentially costly and time-consuming utility relocation. If conflicts cannot be avoided, relocation plans should be prepared. Relocation of the utilities can be done in advance of the project or as part of the project and can be of a temporary or permanent nature. The utility owner must be engaged as early as possible for input and review. Utilities should be shown on the contract drawings. Information such as utility quality level, diameter, depth, material type, owner, trenched cross section, and backfill material used (per ASCE/CI 38) should be identified if available.

Utilities that will remain in place and are in close proximity to the project may require protection and should be monitored during construction. Evaluation of settlement and heave risks may be appropriate depending on type and age of the utility, clearances, and ground conditions.

## 4.4 TRAFFIC FLOW AND ACCESS FOR VEHICLES AND PEDESTRIANS

It is important to collect and evaluate traffic information for the planned alignment as early as possible during planning to determine the feasibility of implementing pilot tube and other guided boring method, to develop alignment alternatives, to plan shaft locations, staging areas, and drive lengths, and to develop means to minimize disruption to vehicles and pedestrians. These traffic data may already exist for the alignments through the city or local municipality. In other cases, it will be necessary to plan and implement a project-specific study. The extent of the data requirements is largely dependent on the nature, complexity, and setting of the proposed construction. The presence of individual facilities with unique access demands, such as schools, post offices, distribution centers, and bus depots must also be considered.

During planning and route selection the alternatives analysis should include the cost and noncost effects to traffic. Cost impacts include traffic control and enforcement as well as reduced delay and fewer miles to travel. Alternative traffic management strategies through work zones may have substantial project cost consequences. For example, if a given method involves time restrictions on construction activities, such as construction during off-peak hours only, the effective workday may be reduced significantly. Thus, the overall duration of project construction increases, substantially affecting project costs. Similarly, other requirements may influence the available work area during certain phases of construction, such as work space limitations that may be due to parallel or adjacent construction activities with potential increased project cost consequences.

In this context, roadway occupancy refers to the degree to which the roadway is *occupied* by construction activities, and, therefore, unavailable for normal traffic flow. As different types of work zone activities occupy varying amounts of the roadway, they also have varying impacts on capacity of the facility.

For most projects, traffic impacts cannot be completely eliminated. Even when the route is fixed, the selection of shaft locations and work spaces can have a significant impact on traffic. Although the individual project considerations vary considerably, the following presents general guidance on shaft siting.

Although most common, the least desirable shaft location is typically within the traffic lanes at an intersection of two roadways. This location could interfere with vehicle turning movements through the intersection as well as movements that cause delays in traffic flow. Another undesirable location would be where activities completely occupy the width of the roadway, thus closing all vehicle traffic from the roadway. Roadway occupancy within 50 ft (15.2 m) of an intersection is also undesirable. This location, depending on the direction of vehicle traffic flow as well as which side of the roadway is occupied, can interfere with turning movements from and through the nearby intersection, thus causing delay to vehicle traffic flow.

A more desirable location for roadway occupancy would be an area at least 100 ft (30.5 m) from an intersection and occupying the middle of the roadway. This location would not, in general, interfere with turning movements from and through the nearby intersection. In addition, vehicle thru-traffic on the roadway could squeeze by the construction activities area with minimal delay to traffic flow. If possible, a roadway occupancy midblock between two intersections, and to one side of the roadway, would be even more preferable. This would allow vehicle traffic to occupy most of the roadway with little or no disruption to vehicle traffic.

## 4.5 ENVIRONMENTAL CONDITIONS

Environmental conditions can be naturally occurring or the result of human activity. Some naturally occurring conditions include naturally occurring hydrocarbons and asbestos as well as active or dormant landslides, active erosional features, and corrosive soils. Some human-created conditions include places of historical and cultural significance as well as landfills and contaminated ground and groundwater. Environmentally sensitive areas can include

- Sources of water supply;
- Islands and island corridors;
- Beaches;
- Dunes;
- Wetlands and wetland transition areas;
- Breeding, nesting, and spawning areas;
- Migratory stopover areas;
- Wintering areas;
- Prime fishing areas;
- Migratory pathways;
- Water areas supporting submerged vegetation;
- Shellfish harvesting waters;
- Forest areas;
- Habitat for federal and state endangered or threatened plant and animal species;
- Federal and state wilderness areas; and
- Areas designated as wild, scenic, recreational, or developed recreational rivers.

Properly planned pilot tube and other guided boring methods can be used to reduce or eliminate the effect of construction on these environmental conditions by undercrossing the features.

Existing site features that could be affected by shaft construction and pilot tube and other guided bore operations should be identified during the site investigation. Historic buildings and environmentally sensitive areas usually require evaluation on a case-by-case basis to ensure protection.

Effects to the surface or the near-surface environment can be reduced by locating jacking and receiving shafts outside the sensitive area and tunneling under it. In this case, the effects are avoided by simply passing under the sensitive area. If the distance is too great for a single drive, then consider placing a receiving shaft at the midpoint and boring to the receiving shaft. This reduces the impact as receiving shafts are smaller than jacking shafts and require less equipment in support of the receiving shaft operation. However, the volume of spoils is increased by the volume of the shaft. Alternatively, consider increasing the bore diameter to achieve a longer drive distance. In this case, the excavated volume is increased based on the increased tunnel diameter and not by an additional shaft.

For contaminated ground, if the affected area cannot be avoided during planning-level route selection, the volume of contaminated spoils generated can be reduced by locating the shafts outside the contaminated area and then advancing the bore though the affected area. In this case, the volume of contaminated ground is reduced to the excavated bore volume. If the distance is too great for a single drive, then consider placing a receiving shaft at the midpoint and advancing to the receiving shaft from both ends. Although adding a receiving shaft increases the volume of spoils generated, it is typically less than with open trench construction. Constructing a receiving shaft creates less volume than a jacking shaft. Alternatively, consider increasing the bore diameter to achieve a longer drive distance. In this case, the excavated volume is increased based upon the increased bore diameter, which may be less than the combined volume of shafts and tunnel. When excavating contaminated ground, always consider the total excavated volume (shafts and tunnels) in arriving at a final design.

Consult an environmental professional for assistance in development of contract requirements related to such issues as handling and disposing of the spoils, decontamination of equipment before and after construction, and personnel health and safety. Note that the availability of contractors qualified to perform work at contaminated sites may be reduced.

## 4.6 FLOOD ZONES

Floods, storm surges, and flash flood conditions should be considered when identifying and selecting shaft locations and tunnel alignments during design. Designing for potential high-water conditions can minimize flooding of shafts, reduce damage to pilot tube and other guided boring equipment, prevent pollution from petroleum products, mitigate cost and schedule impacts, and protect personnel. In addition, locating work areas and shafts within designated flood zones may require additional permitting effort, which can delay design activities.

Assessment of flooding risk during planning should include review of information from FEMA, the National Oceanic and Atmospheric Administration, and United States Army Corps of Engineers as well as local knowledge to determine if storm impacts are probable, along with expected seasonality and frequency of storm occurrence. In some locations, flooding may last for weeks and affect the project duration. The ground conditions may become saturated and can affect the shafts and boring operation.

If flooding conditions are a significant risk, the design may need to consider the following:

- Design shafts rather than the full hydrostatic pressure condition (i.e., increase pressure to the top of the shaft or plan to flood the shaft to eliminate unbalanced hydrostatic pressures).
- Design for fully saturated ground conditions.
- Temporarily raise the site grade, if permittable, to provide better access.
- Design for increased uplift that may occur during flooding or determine maximum groundwater level and method of monitoring that is allowed before flooding occurs.
- Increase pipe joint pressure ratings to accommodate installation and operational high water pressures due to flooding. Design may need to consider 10-, 50-, 100-, or even 500-year flood frequency events.
- Develop plans for evacuation and note that access roads may flood before the project site floods.
- Protect or remove sensitive equipment, oils, and fuels.
- Develop plans to protect the project site from flooding when not occupied, including weekends and holidays.
- Develop plans to monitor water levels in advance and during large rainstorm events.
- Design shafts, pipes, and operations that prevent destabilization of flood control levees and embankments. Recognize that localized levee slope failures can occur during floods.
- Plan construction operations to avoid periods with a high potential for flooding (hurricane season, rainy season).
- Develop start-up plans for after a flood event. Plans should include mobilization of pumps; disposal of water from the shaft; and cleanup of the surface, shaft walls, and equipment within the shaft.

In areas prone to flooding, the owner should consider including allowances or contingencies in the contract to compensate the contractor for this event, associated cleanup, and loss of time to the construction schedule. A major flood event is typically considered a differing site condition. It is recommended to define flood events in reference to specific water surface elevation and recurrence intervals (e.g., 1 in 100 years) and to identify the anticipated conditions in the contract documents.

## 4.7 SEISMIC CONSIDERATIONS

Seismic conditions, including ground shaking and fault offset and related ground responses, should be considered when selecting shaft locations, pipeline alignments, and design requirements for the jacking and receiving shafts. Equally important are the anticipated post-seismic-event level of service and the consequences of failure of the pipeline (e.g., flooding or gas release and fire). The seismic considerations include protection of personnel and minimization of damage from fault offset, ground shaking, landslide, liquefaction, lateral spreading, and tsunamis.

Assessment of seismic risk should include review of information from the U.S. Geological Survey, applicable state geological surveys and departments of transportation, and local knowledge to determine the location, nature, extent, recurrence interval, and design magnitude of seismic conditions relative to the project. Seismic events are of relatively short duration, but their effects can severely alter ground conditions at shaft locations and along the bored alignments.

If seismic conditions are a significant risk, the design should consider the following:

- For areas prone to tsunamis, flooding, as described in Section 4.6.
- Automatic seismic shut-off valves on piping.
- Pipe materials, pipe joints, and joint restraint.
- Reduction of hard points in the alignment (e.g., sharp bends, rigid structure connections, concrete backfill, and entry and exit points of temporary shafts).
- Damage evaluation and start-up plans after a seismic event.
- Ground improvement for critical lifeline or hazardous utilities in areas prone to liquefaction or seismically induced landslides and lateral spreading, including those required by local codes.
- Local code requirements for setbacks from and/or crossing requirements of active faults.
- At fault crossings, the anticipated direction, style, recurrence interval, and displacement of a design seismic event.
- Fault crossing direction with respect to anticipated style and sense of fault displacement (the tensile versus compressive strength of the same material will vary).
- Oversized casing around carrier pipe to allow fault offset/creep deformation to be accommodated by the casing and annular space (air or crushable material) between casing and carrier pipe.
- Tie-in valving to be outside the limits of possible fault movement to allow for quick temporary overland bypassing of fault offset damaged utilities.

In areas prone to seismic events, the engineer should consider the need for ground improvement and that the project could be affected and delayed if an earthquake occurs during construction.

For additional information on seismic design of pipelines, see O'Rourke (2003) and the Technical Council on Lifeline Earthquake Engineering (TCLEE) Publications and Monographs, American Society of Civil Engineers.

## CONSENSUS STANDARDS

ASTM International D1586. "Standard test method for standard penetration test (SPT) and split-barrel sampling of soils." West Conshohocken, PA.

ASTM D3550. "Standard practice for thick wall, Ring-Lined, split barrel, drive sampling of soils." West Conshohocken, PA.

## REFERENCES

ASCE. (2002). "Standard guidelines for the collection and depiction of existing subsurface utility data." ASCE 38-02, Reston, VA, 32.

ASCE. (2007). "Geotechnical baseline reports for underground construction: Suggested guidelines." Technical Committee for the Underground Technology Research Council, Reston, VA.

Barton, N., Lien, R., and Lunde, J. (1974). "Estimation of support requirements for underground excavations." *Proc., 16th Symp. on Rock Mechanics*, ASCE, New York, 163–177.

Bieniawski, Z. T. (1974). "Geomechanics classification of rock masses and its application." *Tunnelling, Proc., 3rd Int. Congress on Rock Mechanics*, Vol. IIA, ISRM, Denver, 27–32.

FHWA (Federal Highway Administration). (2001). "Manual on Subsurface Investigations." National Highway Institute Publication No. FHWA NH 1-01-031, Washington, DC.

Heuer, R. (1974). "Important ground parameters in soft ground tunneling." *Proc., Specialty Conf. on Subsurface Exploration for Underground Excavation and Heavy Construction*, ASCE, Reston, VA, 41–55.

Hunt, S. W., Del Nero, D. E., and Finney, A. J. (2013). "Microtunneling in gravel, cobbles and boulders." *Proc., Rapid Excavation and Tunneling Conf. (RETC)*, SME, Washington, DC.

Kulhawy, F. H., Trautmann, C. H., and O'Rouke, T. D. (1991). "The soil-rock boundary: what is it and where is it?" *Detection of and construction at the soil/rock interface*, ASCE, New York, 1–15.

O'Rourke, M. J. (2003). "Buried pipelines." *Earthquake engineering handbook*, C. Scawthorn and W.-F. Chen, CRC Press, Boca Raton, FL.

Parker, H. W. (2004). "Planning and site investigation in tunnelling." *1º Congresso Brasileiro de Túneis e Estruturas Subterrâneas Seminário Internacional South American Tunnelling*, ITA/AITES.

Sabatini, P. J., Bachus, R. C., Mayne, P. W., Schneider, J. A., Zettler, T. E. (2002). "Geotechnical engineering circular no. 5. evaluation of soil and rock properties." Rep. No. FHWA-IF-02-034, Federal Highway Administration, Washington, DC.

Terzaghi, K. (1950). "Geologic aspects of soft ground tunneling." *Applied sedimentation*, P. D. Trask, ed., Wiley, New York.

# CHAPTER 5
# SHAFT DESIGN

## 5.1 JACKING AND RECEIVING SHAFTS

A pilot tube or other guided boring operation generally requires the design and construction of two shafts: a jacking shaft and a receiving shaft. Refer to Section 3.1.4 for a thorough discussion of the planning process for locating these shafts and providing sufficient work space at each.

In most cases, selection of the shaft support system will be at the discretion of the contractor, subject to performance criteria established in the contract documents. The shafts may need to have an engineered earth support system if space is limited or the shafts are deep. In such cases, the engineered support system should be specific to the site, prepared by a licensed engineer, and submitted for review to the owner's representative. Depending on the shaft depth and available work space, sloping of the shaft walls may be sufficient. Regardless of whether the shaft has an earth support system, all elements of shaft design and construction must comply with Occupational Safety and Health Administration standards. Available shoring methods may include: trench boxes, slide rail systems, steel frames and plates, and corrugated pipes. Engineered earth support systems may consist of soldier piles and lagging, liner plates, precast concrete rings, precast concrete caisson, and steel sheet piles. Other earth support systems may be used if designed by a licensed engineer.

When designing the overall length of the jacking shaft, the shaft design engineer should consider the length of the jacking frame with extensions; the thrust block and entry seal if used; and the length of the jacking pipe, pilot tube, or thrust casing. It also should be large enough to accommodate the boring equipment plus room for the operator; additional workers for adding pilot tubes, thrust casings, and jacking pipes; means of egress; and other worker safety requirements. The additional shaft depth below the pipe invert should consider: height of the jacking frame; depth of the skip pan; welding of casing or carrier pipe, if required; and water control measures.

Once the shaft excavation is complete, a stable subgrade and working surface must be built in the shaft bottom. This stable surface can be achieved by using crushed stone, gravel, steel plates, wood beams, concrete mud slab, reinforced concrete, or a combination thereof as geotechnical considerations warrant for the project. The floor of the shaft is set to the proposed slope of the bore unless the jacking equipment has grade adjustment capability.

The jacking equipment may require a thrust reaction system to offset the thrust forces needed to advance the pilots, casings, and pipe string. The size and components of the reaction system will vary depending on the excavation support system, equipment size, and the anticipated jacking forces required, plus a suitable safety factor. It may be feasible to react against the back of the pit, provided the excavation support system will not be compromised. In some cases, the excavation support system may need to be strengthened to receive the anticipated loads.

The effect of the groundwater elevation on shaft design and construction should be considered. Dewatering efforts may be required to lower the groundwater at the shafts as well as along the alignment. Refer to Chapter 4 for further discussion on groundwater. The contractor must be prepared to accommodate the dewatering equipment during pilot tube and other guided boring operations and shaft construction. As is the case with every dewatering system, the contractor should take measures to prevent migration of ground. During dewatering operations, treatment and discharge plans should be developed that comply with regulatory agency requirements.

Surface water should be managed so that the shaft subgrade remains dry and stable. This can be accomplished using ditches and/or berms to divert the overland flow around and/or away from the shaft. A sump pump may be needed to handle rainwater in the shaft.

## 5.2 LOCATION

Jacking and receiving shaft locations should be selected to minimize the number of shafts and disruption, and, at the same time, to not exceed the practical limits of drive lengths achievable for a given set of site conditions and constraints. Maximum drive length will be a function of boring equipment used, diameter of the pipeline, jacking strength of the selected pipe, ground conditions, lubrication, and pipeline maintenance requirements. Project risk increases when longer drives are used.

Ideally, the shafts should be planned and built at manholes or lateral connections in sewer lines, at fire hydrants, valves or bends in pressure applications, and at splice points for high-voltage cables. One strategy for minimizing the number of jacking shafts is to drive in two or more directions from a single shaft.

Shaft locations should be selected to minimize impact to traffic and businesses and to avoid site features that could result in inefficiencies during construction such as low overhead utilities and/or underground utilities that cannot be relocated. Limitations on staging and laydown areas at shaft locations, work hour restrictions, and size limitations of the shafts must be clearly identified in the contract documents.

Receiving shafts are typically smaller, require less working space, and are occupied for shorter duration than jacking shafts. Accordingly, a receiving shaft is the better choice when a shaft must be located in areas with limited space, or where surface disruption needs to be kept to a minimum.

Some flexibility in shaft locations may be needed as a contractor may request alternate locations. The contractor may need some additional freedom with shaft locations to accommodate its ground support system and equipment while minimizing the impact to utilities and other adjacent structures. Alternate shaft locations will require approval by the project owner.

## 5.3 SHAFT DESIGN CONSIDERATIONS

Design criteria and performance requirements for shafts should be provided in the contract documents. As applicable, the following minimum design criteria and performance requirements should be included:

- Applicable codes and standards;
- Groundwater control and disposal constraints;
- Allowable settlement of the ground surface and structures adjacent to the shafts;
- Acceptable excavation support wall system types;
- Limitations on the use of external lateral support methods;
- Constraints and requirements relative to relocation of existing utilities within the footprint of the shafts; and
- Methods for exit and entry of the boring equipment.

Engineered earth support systems must be designed by or under the direct supervision of a registered professional engineer licensed in the state where the project is located. The design must be signed and stamped by that professional engineer.

In addition to the minimum design criteria and performance requirements listed, additional information should be provided by the engineer to be used by the shaft designer. This information includes

- Soil, rock, groundwater, and minimum surcharge loading;
- Minimum factors of safety relative to bottom stability;
- Unacceptable excavation support wall system types; and
- Parameters for thrust wall design.

The use or non-use of a dewatering system will control which ground support methods can be effectively used. In some cases, contaminated groundwater will preclude the use of dewatering wells. Restricting groundwater disposal quantities, limiting discharge locations, and requiring special testing and water treatment may also preclude the use of dewatering systems.

In some ground conditions, removal of the shaft support system may displace adjacent ground, resulting in settlement of manholes or pipe constructed inside the shaft and/or adjacent utilities. Some owners and local municipalities require partial or complete shoring removal. If removal is required, the contract should specify that the shaft support system be removed in a manner that minimizes ground movement.

## 5.4 TRENCHLESS ENTRY/EXIT FROM SHAFTS

To implement pilot tube and other guided boring operations, openings must be created in the earth support system to allow for the installation and retrieval of the pilot tubes, thrust casings, and jacking pipes. Shaft designs and work plans must include provisions for portal stability at these shaft penetrations to prevent inflows of ground and lubricants during all stages of the pilot tube and other guided boring operation. Shaft penetrations need to accommodate different size openings, including the installation of pilot tubes, thrust casings, and jacking pipe within the jacking shaft, as well as during pilot tube and thrust casing retrieval at the receiving shaft.

## 5.5 THRUST BLOCKS

During pilot tube advance or guided boring operations, the jacking thrust is usually transferred to the back of the shaft where it is resisted by the passive pressure of the ground behind the shaft. A supplemental backing plate or thrust (reaction) block may be installed against the rear wall of the shaft to uniformly distribute the thrust force exerted by the jacking equipment to prevent shaft damage and misalignment of the jacking frame. The backing plate may be steel plate, structural steel, timber, concrete thrust block, or a combination thereof.

The thrust force exerted by the jacking system and distributed through the backing plate and thrust block (if used) must not exceed the passive pressure of the ground immediately behind the shaft with an acceptable factor of safety. If the jacking shaft is located in ground unable to provide the required thrust restraint, other means of thrust restraint must be employed such as: ground improvement, steel sheets driven into the ground, bolts anchored into the concrete floor, driven steel piles, or a wall constructed integral with a concrete floor extending below grade. The need for ground improvement or other measures should be considered during the design phase and included in the contract documents.

Adequate care should be taken to ensure that the thrust pressure developed by the operation does not affect existing utilities near the jacking shaft or adjacent surface grades.

## 5.6 COMMON SHAFTS

Shafts used by multiple contractors should be identified and properly designed so that joint-use issues will be properly identified and resolved.

Common shafts are defined as shafts that may be used by two different contractors simultaneously or sequentially. They are typically used between adjacent construction contracts. The use of common shafts can result in disputes between contractors that can ultimately have detrimental impact on the project. Therefore, common shafts should be avoided, when possible. If common shafts cannot be avoided, the contract documents must contain a mechanism to minimize the potential for disputes between contractors and a means for resolving such disputes if they occur.

# CHAPTER 6
# PIPE CHARACTERISTICS AND DESIGN

## 6.1 GENERAL REQUIREMENTS

Pipe used in pilot tube and other guided boring method installations should meet the following general requirements:

- Circular shape with a flush outside surface (including the joints).
- Strength sufficient to withstand both the installation loads and the in-place, long-term service loads.
- Dimensional tolerances on length, straightness, roundness, end squareness, and allowable angular deflection.
- Durability for the service exposure (e.g., internal and external corrosion resistance, resistance to internal abrasion).
- Joints capable of watertight performance for internal operating pressures (including working and transient pressures) as stated in the purchaser's documents, and external head pressures resulting from groundwater and lubrication. The joint should also be capable of efficiently transferring jacking forces from the hydraulic jacking frame, through the pipe wall and joint, to the pilot tube or other guided boring equipment.

## 6.2 MATERIAL TYPES

Most pipes installed into the ground by the pilot tube and other guided boring methods are pushed, or direct jacked, into place using either a one-pass or two-pass process. Because pilot tubes and other reaming tools create a smaller-diameter hole in the ground and are able to keep the hole supported at all times, the process also allows small-diameter pipe, or pipe bundles, to be pulled in tension back into the hole. Pipe materials currently used with the pilot tube and other guided boring methods in direct jacked or pulled back applications are (listed alphabetically):

- Direct jacked pipe materials:
  - ductile iron (DI),
  - fiberglass-reinforced polymer mortar (FRPM),
  - polymer concrete (PC),
  - polyvinyl chloride (PVC),
  - reinforced concrete (RC),
  - steel, and
  - vitrified clay (VC).
- Pulled back pipe materials:
  - high density polyethylene (HDPE),
  - polyvinyl chloride (PVC), and
  - steel.

Pilot tube and other guided boring methods are widely used for sewer and stormwater applications. The process involves the direct jacking of pipe to advance the boring equipment through the in situ ground conditions. Low-head, or gravity, pipes most often jacked in these nonpressure applications include these types: FRPM, PC, RC, and VC. All of these pipe materials have a substantial installation history in sewer applications.

The use of pilot tube and other guided boring methods for direct jacking installation of pressure pipes has been limited because such installations typically follow the general topography along the pipeline alignment, with a minimum cover. These installations do not have the precise grade requirements that are needed for a gravity system. However, when used for pressure pipe installations, suitable pipe materials include DI, FRPM, PVC, and steel.

Pilot tube and other guided boring methods are routinely used to install steel pipe as a casing in a two-pass installation process. In the first pass, the steel casing is jacked into the ground. Other jacking pipe materials can be used as a casing, although the thicker sidewall reduces the available internal volume for a carrier pipe. In the second pass, the carrier pipe is inserted into the casing. The carrier pipe can be of any pipe material, with a variety of joint configurations; although a flush wall pipe increases the ability to realign the carrier pipe to design line and grade. For more information on carrier pipe materials, pipe joints, and jointing options, see *Manual and Reports on Engineering Practice (MOP) No. 125* (ASCE 2013).

### 6.2.1 Pipe Material Selection

Pipe material selection should be based on many factors, including the following:

- Pipeline operating conditions (pressure: operating, test, transient, and vacuum);
- Pipeline service environment (fluid, temperature, corrosivity, and soil or groundwater contamination);
- External loads (earth loads, surface live loads, and groundwater head);
- Inside diameter of the pipe required by the owner;
- Compatibility with the pilot tube tooling (outside diameter) and jacking system;
- Available work space, shaft size, and jacking frame (pipe section length);
- Anticipated jacking or pulling forces, and drive/pull lengths;
- Pipe deformation, rebound (during jacking), and contraction (after pulling) for plastic and elastic materials;
- Pipe hydraulic characteristics;
- Thermal and electrical conductivity characteristics;
- Pipe availability, reliability, and durability; and
- Lifecycle cost.

When external loads are estimated, arching of the ground should be considered. Trench load factors calculated with such approaches as Marston's equation should not be used to estimate external loads on direct jacked or pulled back pipe. As stated in MOP 60/WEF FD-5 (ASCE, 2007), when Marston's equation is modified to include cohesion, it may be used to determine the gravity soil loads on direct jacked or pulled back pipe through undisturbed soil.

### 6.2.2 Reference Product Standards

Tables 6-1 and 6-2 provide standards for the different direct jacked and pulled back pipe materials.

The reference standard for RC pipe is ASTM C76. ASTM C76 is written for concrete pipe used for open-trench construction, and the tolerances and joint design are not appropriate for direct jacked pipe. See Section 6.3.3 for recommended tolerances for all direct jacked pipe. See Section 6.3.5 for recommended pipe joints.

### 6.2.3 Scope and Range

Tables 6-3 and 6-4 provide size ranges for different direct jacked and pulled back pipe materials. Some of the larger pipe sizes are only applicable to the other trenchless methods discussed in Chapter 1.

Table 6-1. Standards for Different Direct Jacked Pipe Materials

| | Applicable Standards | |
|---|---|---|
| Material Type | Nonpressure | Pressure |
| DI | AWWA C150/C151<br>ASTM A716<br>ASTM A746 | AWWA C150/C151 |
| FRPM | ASTM D3262 | ASTM D3517<br>ASTM D3754<br>AWWA C950 |
| PC | DIN 54815-1 & 2<br>ASTM D6783<br>DIN EN 14636-1 | N/A |
| PVC | ASTM D3034 | ASTM D1785<br>ASTM D2241<br>AWWA/ANSI C900<br>AWWA/ANSI C905 |
| RC | ASTM C76<br>ASCE 27-00 | ASTM C361<br>AWWA C300/C302 |
| Steel | ASTM A36<br>ASTM A139 Grade B[a]<br>ASTM A515<br>ASTM A572<br>API 2B[b] | AWWA C200<br><br><br><br>API 2B |
| VC | ASTM C1208<br>DIN EN 295-7 | N/A |

Note: N/A = not applicable.
[a]No hydrostatic test required.
[b]Dimensional tolerances only.

Manufacturers and pipe suppliers should be consulted for the latest size ranges, application, availability, and suitability. Units used are based on the most common pipe standards. Refer to the specific pipe standards for information on standard units of manufacture.

Actual sizes of pipe available and applicable to pilot tube and other guided boring methods depend on current state of the art at the time of design and construction. Actual internal diameters and external diameters may vary from the nominal. Consult the various pipe manufacturers for applicable diameters and wall thicknesses.

Table 6-2. Standards for Different Pulled Back Pipe Materials

| | Applicable Standards | |
|---|---|---|
| Material Type | Nonpressure | Pressure |
| HDPE | ASTM F714<br>AWWA/ANSI C906 | ASTM F714<br>AWWA/ANSI C906 |
| PVC | ASTM D3034<br>ASTM F679 | ASTM D1785<br>ASTM D2241<br>AWWA/ANSI C900<br>AWWA/ANSI C905 |
| Steel | ASTM A36<br>ASTM A139 Grade B[a]<br>ASTM A515<br>ASTM A572<br>API 2B[b] | AWWA C200<br><br><br><br>API 2B |

[a]No hydrostatic test required.
[b]Dimensional tolerances only.

Table 6-3. Size Ranges for Different Direct Jacked Pipe Materials

| | | Operating Condition | |
|---|---|---|---|
| Material Type | Range of Nominal Diameters (in.) | Nonpressure | Pressure |
| DI | 4 to 16 | Yes | Yes |
| FRPM | 18 to 126 | Yes | Yes |
| PC | 8 to 102 | Yes | N/A |
| PVC | 8 to 15 | Yes | Yes |
| RC | 12 to 144+ | Yes | N/A |
| Steel | 4 to 144+ | Yes | Yes |
| VC | 8 to 42 | Yes | N/A |

Note: N/A = not applicable.

### 6.2.4 Section Lengths

Table 6-5 provides typical lengths of pipe sections for the different direct jacked and pulled back pipe materials. The lengths of pipe sections are presented in U.S. customary units, when applicable, because they are manufactured products with dimensions not converted to International System (SI) units. The length of the pipe used on a project will be a function of the shaft or pit size, equipment used, available work space, and access.

Table 6-4. Size Ranges for Different Pulled Back Pipe Materials

| Material Type | Range of Nominal Diameters (in.)[a] | Operating Condition | |
|---|---|---|---|
| | | Nonpressure | Pressure |
| HDPE | 4 to 65 | Yes | Yes |
| PVC | 4 to 48 | Yes | Yes |
| PVC: Fusible | 4 to 36 | Yes | Yes |
| Steel | 4 to 144+ | Yes | Yes |

[a]Pipe used in pulled back installations (individual or bundled) are typically 16 in. and smaller.

Table 6-5. Typical Pipe Lengths

| Material Type | Standard Lengths | Available Lengths |
|---|---|---|
| DI | 19.5 ft | As required |
| FRPM | 10 ft/20 ft | 4 ft/5 ft/6.5 ft/8 ft |
| HDPE | 40 ft/45 ft | As required |
| PC | 8 ft/10 ft and 1 m/2 m/3 m | 3 ft/6 ft |
| PVC | 2 ft/4 ft/6 ft | As required |
| PVC: Fusible | 40 ft/45 ft | As required |
| RC | 7.5 ft to 24 ft | Varies |
| Steel | 8 ft/10 ft/20 ft/40 ft/60 ft | Any |
| VC | 4 ft/6 ft/8 ft/10 ft and 1 m/2 m/3 m | >2 ft |

For the HDPE and PVC pulled back pipe, consideration must be given to the size of the shaft or pit where the material is being pulled from. The work area must be large enough either to accommodate the butt fusion welding equipment or to allow for bending of the pulled back pipe material from the surface. For the smaller-diameter HDPE pipe, the material may be available on a continuous spool, which would be placed in the shaft.

### 6.2.5 Mechanical Properties

All pipe materials used will need to have sufficient mechanical strength to provide the required external and internal load-bearing capability for the installation and application required. The installation will require that the pipe have either sufficient compressive capacity for a pipe jacking operation, or sufficient tensile capacity for a pipe pulling operation.

During installation of pipe through ground conditions with gravel, cobbles, or boulders, rock pieces can enter into the annular space and point load the direct jacked or pulled back pipe. The flexural bending and tensile strength of the pipe material are important mechanical properties to evaluate when selecting the pipe for these complex ground conditions.

### 6.2.6 Manufacturing and Quality Control

The pipe manufacturing process and quality control must ensure that the required characteristics are achieved, including

- General workmanship,
- Strengths and loading capacities,
- Dimensions and tolerances,
- Joint integrity, and
- Surface consistency (smoothness and linings or coatings, if any).

Specific procedures and test methods vary with jacking pipe materials. Additional needs for the pipe at the time of manufacture, such as coatings and linings, should also be specified per the specific installation and project constraints as well as applicable code and jurisdictional requirements.

## 6.3 PIPE DESIGN

### 6.3.1 In-Use Service Loads

The pipes and joints, once installed, must be capable of safely withstanding, for the project design life, all in-use service loads, including

- Internal operating, transient, and test pressures;
- Earth overburden (trenchless construction may result in significantly lower loads than those experienced with open-cut installations);
- Surface loads (i.e., surcharge, vehicle, railroad, and/or aircraft, when present);
- External static water head and vacuum (if present);
- Thermal requirements;
- Vibrations;
- Corrosion and chemical deterioration; and
- Abrasion associated with periodic cleaning, if applicable.

Design procedures for these conditions vary somewhat by pipe material and are detailed in many other documents and standards. Therefore, they are not addressed further in this manual. Begin with product-specific design manuals and standards, and consult the pipe manufacturers for more details.

### 6.3.2 Installation Loads

**6.3.2.1 Jacking Forces.** Jacking pipe is subject to transient axial loads, called jacking forces, applied during the installation process. Depending on the pilot tube or other guided boring methods being used, the jacking forces may include the force required to advance the pipe and the auger casings as well as the force associated with advancing the upsizing tools in the horizontal direction.

Jacking forces are greatest on sections of pipe nearest the jacking shafts or directly in front of the jacks. These forces are rarely distributed evenly around the pipe's end circumference because the squareness and mating of joints and pipe alignment are seldom perfect. Some eccentricity of the axial load typically occurs in the field. The result is force concentrations in portions of the pipe ends (joints).

The maximum jacking stress at any point in the pipe string is at least as great as the maximum jacking force recorded at the jacks (minus any friction losses along the drive between the jacking shaft and the point in question) divided by the effective minimum cross-sectional area of the pipe wall.

The pipe and joints must be able to withstand these stresses without cracking, breaking, or suffering other damage. Frequently, pipe ends of segmented pipes are separated with compressible packers (also called compression rings or cushion rings). These compression rings are normally used to distribute jacking forces between the pipe ends more evenly. The cushion material selected, and its performance under repeated loads, may cause hardening of the material during the installation. Most compression ring materials selected have a low Poisson's ratio. Steel casings are often welded or have rigid mechanical designs that do not accommodate compression rings.

As the pipes are jacked, the pipe and joint exterior surfaces will experience skin friction from the surrounding ground. The pipes and joints must have sufficient durability and toughness to withstand this phenomenon without significant abrasion, loss of joint seal, damage, loss of corrosion resistance, or failure. Past successful experience in similar conditions is typically the best evidence of satisfactory capabilities. Adequate overcut, annular thickness, and lubrication can significantly reduce skin friction.

The factor of safety for jacking forces is the ratio of the pipe's ultimate (or yield) axial compressive capacity divided by actual applied load at the jacks. The allowable load calculated using the factor of safety might need to be reduced because of eccentric loading or end squareness tolerances. Different pipe materials require different safety factors to allow for various material characteristics.

**6.2.3.2 Pullback Forces.** Pullback installation of fused, continuous, or restrained joint pipe will subject the pipe to axial tensile forces. These forces are generated primarily from the friction of pulling the pipe back through the excavated bore; however, forces may also be generated by bending of the pipe as it is pulled into the required alignment from grade, if applicable. The pulling forces can be reduced with adequate overcut, annular thickness, and the use of lubrication.

The pulling forces are frequently largest on sections of pipe nearest the pulling head. Similar to the direct jacked installation, pipe tensile stress may also be highest at the pulling head but may exceed these values locally because of point loading, bending, or eccentricity.

The pulled back pipe and joint must have the tensile capacity and durability to withstand the installation loading without significant abrasion, loss of joint seal, damage, loss of corrosion resistance, or failure. Additionally, if bending will be required to insert the pulled back pipe into the installation, the pipe and joint should be of sufficient strength to withstand the bending stresses required.

**6.2.3.3 External Pressure.** External pressure is exerted by groundwater and earth loads. Lubrication for the annular space can be injected under pressure from a power cutter head, through the pilot tube equipment, and/or through ports in the pipe walls, and can impose an additional external pressure. The pipes and joints must not leak, be damaged, or fail from applications of these pressures.

### 6.3.3 Dimensional Criteria

The engineer should provide tolerances for each of the following dimensional characteristics parameters for the type of pipe material included in the contract documents:

- Circumference uniformity,
- Exterior roundness,
- End squareness or planeness,
- Section straightness,
- Pipe length, and
- Allowable angular joint deflection tolerance.

These six tolerance parameters can have a significant effect on the jacking process and pipe performance. These values are used by the contractor to determine the required bore diameter and the associated overcut.

The following sections explain the importance of each of the dimensional criteria for these parameters as it relates to a successful pilot tube and other guided boring methods installation:

*End squareness* is the geometric relationship between the barrel of the pipe and the pushing surfaces of the pipe joint. If the pipe joint face is not perpendicular to the axis of the pipe, the jacking forces are not evenly distributed around the bearing surface, resulting in areas of much higher stress concentrations or point loads. To compensate for some variations in end squareness, pilot tube and other guided boring methods contractors install compression rings or cushion rings that can more effectively redistribute the jacking loads and avoid potential joint damage.

*Straightness* deviation is often stated in standards as a deviation from a straight edge placed longitudinally above the pipe on the concave side. This deviation is reflected on the opposite side of the pipe and will make the effect of the pipe larger than the maximum diameter.

*Eccentricity* from the pipe axis of pipe joints affects the design of the shield cut and required bore of the installation. This eccentricity should be included in the borehole calculation to provide adequate annulus.

*Maximum outside diameter (OD)*, diameter tolerance, ovality, pipe straightness, and pipe eccentricity are all elements that affect the most critical dimension used in pilot tube and other guided boring methods. These values are used by the contractor to determine the required bore diameter and the associated annulus.

Pipe materials vary widely in their resistance to corrosive effects, both internally and externally. The pipe selection should evaluate the need to resist corrosion and the desired life of the installation by the owner. In some materials, internal or external sacrificial material may be added to the wall to extend the life of the material. This material should be allowed for in the design and installation annulus calculations.

The standards of manufacture of pipes vary widely with respect to measurement methods, use of appropriate statistical analysis, and dimensional limits. The engineer and installer must evaluate these variables for successful installation design and pipe selection.

Pilot tube and other guided boring installation methods are limited as to the amount of overcut and shield cut. They require precise confinement of the boring equipment and pipe string and are sensitive to fluctuation in the aforementioned dimensional parameters. The following critical dimensional parameters may have an effect on several aspects of the jacking process.

*Jacking forces.* As variations in circumference, roundness, end squareness or planeness, and section straightness increase, so do the jacking forces, sometimes substantially. This increase can result in decreases in the distance for safe jacking drives and/or in an increased need for intermediate jacking stations; ultimately, in severe cases, this increase can cause pipe failures.

*Pulling forces.* As variations in circumference, roundness, end squareness, and section straightness increase, so do the pulling forces, sometimes substantially. This increase can result in decreases in the safe pull distances, and in severe cases can cause pipe failures.

*Load transfer.* Square, plane pipe ends, and straight sections improve the uniformity of jacking load distribution on the pipe ends.

As deviations of straightness, squareness, and planeness increase, uneven loading on the pipe ends also increases. These load concentrations and point loads, when severe enough, may cause pipe damage or failure, sometimes even at forces below the calculated allowed load on the pipe.

*Summary.* Although no pipes have perfect dimensional control, tolerances on these characteristics vary significantly from one pipe material to another, and from one manufacturer to another within a given type of pipe material. The dimensional tolerances required for successful performance vary according to several factors, including the following:

- Drive or pull distance;
- Magnitude of jacking or pulling forces versus the ultimate pipe load capacity;
- Operational control, such as amount of overcut and effectiveness of lubrication;
- Compounding effects of several dimensional criteria;
- Ground conditions; and
- Type of pipe material.

In spite of the previously noted influences, pipe users generally agree that satisfactory jacking performance at a reasonable cost is achieved when the pipe dimensional tolerances are maintained within certain limits. Table 6-6 provides desirable dimensional tolerances for all pipe types. The engineer should consult with the pipe manufacturers and installer to determine whether appropriate dimensional tolerances can be met.

Tolerances outside of the desirable limits do not automatically ensure that pipe cannot perform successfully; however, it may be necessary to increase the jacking or pulling load safety factor to achieve acceptable reliability, increase the annulus, and/or increase the pipe load rating (capacity) to safely complete the required installation distances.

Table 6-6. Desirable Dimensional Tolerances for All Pipe Material Types

| Dimensional Characteristic | Desirable Tolerance |
|---|---|
| Outside Diameter | Within 0.1% |
| Exterior Roundness | Within 0.5% |
| End Squareness or Planeness | |
| Diameter ≤ 48 in. | Within 1/16 in. |
| Diameter > 48 in. | Within 1/8 in. |
| Straightness | Within 1/8 in. |
| Joint Length | Within 1/4 in. |

### 6.3.4 Pipe Lubrication

Pipes with diameters large enough to permit personnel entry are frequently equipped with lubrication ports (fittings) in the wall, to permit injection of a lubricant during jacking to reduce jacking forces or placement of grout after jacking to fill any residual annular space. Pipes should be equipped with these fittings at the frequency determined by the contractor or required by the engineer. In cases in which drives are short and the ground friction is expected to be low, with resulting low jacking forces, the engineer should consider whether to exclude the lubrication ports. Some pilot tube and other guided boring equipment is capable of injecting lubricant with the upsizing tools, without personnel entry.

### 6.3.5 Pipe Joints

**6.3.5.1 Direct Jacked Pipe Joints.** All jacking pipe joints shall meet the following performance requirements:

- Flush or slightly recessed in relation to the pipe OD;
- Capable of watertight performance for internal operating pressures (including working and transient pressures), as stated in the contract documents, or external head pressure due to groundwater; and
- Capable of efficiently transferring jacking loads from the hydraulic jacking frame, through the pipe wall and joint, to the boring equipment at the point of excavation.

Most jacking pipe compression joints are gasket sealed to facilitate rapid assembly and to provide the required rotational capabilities and connection-sealing performance. Steel pipes are also available with either interlocking push-together rigid joints (snap joints) or butt-welded rigid joints. There are several gasket-sealed joint configurations. A typical joint configuration is shown in Fig. 6-1.

Compression rings (packers or cushions) made of plywood, particle board, or other low-modulus (compression-resistant) materials should be placed between the pipe ends to reduce load concentrations on the pipes caused by nonplane or nonsquare ends, nonstraight sections, or angles between the axes of adjacent pipes (caused by steering). Compression rings should be as uniform as possible in compression resistance so as not to be the cause of load concentrations.

Gaskets are typically ethylene propylene diene monomer (EPDM), nitrile (buna-N), or neoprene. The gasket material is usually selected based on the service conditions and host environment (including considerations for contaminated ground and groundwater). Other than the gasket material, the installed gasket dimensions must be compatible with the joint type and dimensions to form an effective joint seal.

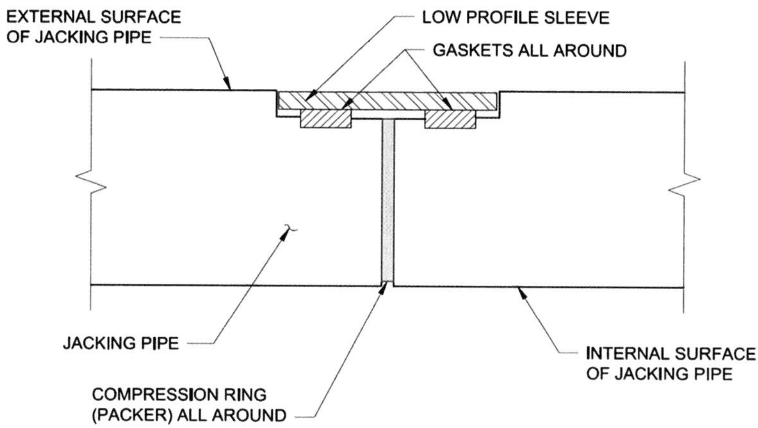


*Fig. 6-1. Typical joint configuration for jacking pipe*

**6.3.5.2 Pulled Back Pipe Joints.** All pulled back pipe joints shall meet the following performance requirements:

- OD capable of insertion into bored hole;
- Capable of withstanding installation pull-back forces without the loss of seal, joint damage, or elongation and necking; and
- Capable of watertight performance for internal operating pressures (including working and transient pressures), as stated in the contract documents, or external head pressure due to groundwater.

Pulled back joint configurations contain either a welded or fused connection, which is similar in size to the pipe barrel, or a mechanical coupler or bell, which provides enough tensile capability for the insertion loading. The pulled back joint type will depend on the pipe material and commercially available joining technologies.

### 6.3.6 Resistance to the Environment

Pipe material and appurtenances and their resistance to environmental factors should be considered in the selection process. The use of coatings and linings is a function of the pipe material selected and of the environment.

Pipes and joints will be exposed in the long term internally to the fluids and/or gases conveyed, and externally to groundwater, chemicals in the ground, and occasionally stray electrical currents and contamination.

Much information is available from other sources regarding corrosive or galvanic attack of the various materials covered in this manual and the methods available for protection, if needed. Pilot tube and other guided boring installation exposes the exterior of the pipe to friction, which may have a damaging effect on the durability of the materials and any coatings, linings, or other protective measures used.

Pipe installed by direct jacking, including steel casings, typically does not have a coating. The coating may be damaged by the jacking process as the exterior of the pipe is pushed through the ground.

Use of linings varies with the type of pipe used. HDPE, PC, PVC, FRPM, and VC are not typically subject to galvanic corrosion. DI, RC, and steel can be prone to corrosion in some environments, and addition of a lining can increase the pipe's effective life. A lining can be added before or after jacking, depending on the pipe inside diameter. Installing a lining on a pipe before jacking runs the risk of damage to the lining.

A casing installed by jacking and then slip lined with a carrier pipe is a typical means of installing a pipe that needs a coating and/or lining. The space between the carrier pipe and casing is filled with a backfill material that may also improve and provide protection.

Because joint bell sleeves may be a different material than that used for the pipe (e.g., steel or stainless steel collars are frequently used with PC, RC, and VC pipes), separate consideration of resistance to the environment is prudent.

Consult with the various manufacturers of the pipe materials for applicable means of resisting the environmental factors anticipated for the project. With all of the pipe materials, the engineer should use a risk-based protocol to determine the best means and methods of addressing how to resist the environment in which the pipe string is to be installed. Giving the contractor as many pipe materials and installation options as possible can be cost-effective for the project.

### 6.3.7 NSF 61 Certification

All pipe materials that will be in contact with potable water should be certified to meet NSF/ANSI 61. Pipes used for sanitary sewer and storm drain installations do not need NSF/ANSI 61 certification.

For potable water application, some of the pipe materials identified (in Section 6.2) have NSF/ANSI 61 certification. Check with the specific pipe manufacturer to determine if the proposed water pipe material has gone through and received NSF/ANSI 61 certification. Certification can take months to obtain and is expensive. Check with the pipe manufacturers early if considering the use of a new pipe material for potable water applications. Certification of a pipe material is dependent on the final product. Pipe materials made from individual components from different sources, like sand and binder, may need recertification if the source of the component changes.

To avoid the need and expense of pipe certification, some potable water installations use the pulled back method or a two-pass jacked installation method. A casing is jacked into the ground using the pilot tube and other guided boring methods, and then the NSF/ANSI 61 certified potable water pipe material is inserted into the casing. The space between the carrier pipe and casing is typically filled with backfill grout.

## CONSENSUS STANDARDS


ASTM International D3754. "Standard specification for "fiberglass" (glass-fiber-reinforced thermosetting-resin) sewer and industrial pressure pipe." West Conshohocken, PA.

ASTM D6783. "Standard specification for polymer concrete pipe." West Conshohocken, PA.

ASTM F714. "Standard specification for polyethylene (PE) plastic pipe (DR-PR) based on outside diameter." West Conshohocken, PA.

ASTM A36/A36M. "Standard specification for carbon structural steel." West Conshohocken, PA.

ASTM A716. "Standard specification for ductile iron culvert pipe." West Conshohocken, PA.

ASTM A746. "Standard specification for ductile iron gravity sewer pipe." West Conshohocken, PA.

ASTM D3517. "Standard specification for "fiberglass" (glass-fiber-reinforced thermosetting-resin) pressure pipe." West Conshohocken, PA.

ASTM A572/A572M. "Standard specification for high-strength low-alloy columbium-vanadium structural steel." West Conshohocken, PA.

ASTM D1785. "Standard specification for poly(vinyl chloride) (PVC) plastic pipe, schedules 40, 80, and 120." West Conshohocken, PA.

ASTM D2241. "Standard specification for poly(vinyl chloride) (PVC) pressure-rated pipe (SDR series)." West Conshohocken, PA.

ASTM A515/A515M. "Standard specification for pressure vessel plates, carbon steel, for intermediate- and higher-temperature service." West Conshohocken, PA.

ASTM A139/A139M. "Standard specification for electric-fusion (arc)-welded steel pipe (NPS 4 and over)." West Conshohocken, PA.

ASTM D3262. "Standard specification for "fiberglass" (glass-fiber-reinforced thermosetting-resin) sewer pipe." West Conshohocken, PA.

ASTM F679. "Standard specification for poly (vinyl chloride) (PVC) large-diameter plastic gravity sewer pipe and fittings." West Conshohocken, PA.

ASTM C76. "Standard specification for reinforced concrete culvert, storm drain, and sewer pipe." West Conshohocken, PA.

ASTM C361M. "Standard specification for reinforced concrete low-head pressure pipe (metric)." West Conshohocken, PA.

ASTM D3034. "Standard specification for type PSM poly(vinyl chloride) (PVC) sewer pipe and fittings." West Conshohocken, PA.

ASTM C1208/C1208M. "Standard specification for vitrified clay pipe and joints for use in microtunneling, sliplining, pipe bursting, and tunnels." West Conshohocken, PA.

AWWA (American Water Works Association) C150. "Thickness design of ductile—Iron pipe." Denver.

AWWA C900. "Polyvinyl chloride fabricate fittings, 4 in. through 12 in. (100 mm through 300 mm), for water transmission and distribution." Denver.

AWWA C151. "Ductile—Iron pipe, centrifugally cast." Denver.

AWWA C905. "Polyvinyl chloride (PVC) pressure pipe and fabricate fittings, 14 in. through 48 in. (350 mm through 1,200 mm), for water transmission and distribution." Denver.

AWWA C302. "Reinforced concrete pressure pipe, noncylinder type." Denver.

AWWA C300. "Reinforced concrete pressure pipe, steel cylinder type." Denver.

AWWA C200. "Steel water pipe 6 in. (150 mm) and larger." Denver.

AWWA C950. "Fiberglass pressure pipe." Denver.

AWWA C906. "Polyethylene (PE) pressure pipe and fittings, 4in. through 65 in. (100 mm through 1,650 mm), for waterworks." Denver.

DIN (Deutches Institut für Normung) 54815-2. "Pipes of filled polyester resin moulding materials. Part 2: Requirements, testings." Berlin, Germany.

DIN 54815–1. "Pipes made of filled polyester resin moulding materials—Dimensions, material and marking." Berlin.

DIN/EN 14636-1. "Plastics piping systems for non-pressure drainage and sewerage—Polyester resin concrete (PRC). Part 1: Pipes and fittings with flexible joints." Berlin.

DIN/EN 295-7. "Vitrified clay pipe system for drains and sewers. Part 7: Requirements for pipes and joints for pipe jacking." Berlin.

## REFERENCES


API (American Petroleum Institute). (2001). "Specification for the fabrication of structural steel pipe." *API Spec 2B*, Washington, DC.

ASCE. (2000). "Standard practice for direct design of precast concrete pipe for jacking in trenchless construction." *ASCE 27*, Reston, VA.

ASCE. (2007). *Gravity sanitary sewer design and construction*, 2nd Ed., ASCE, Reston, VA, 436.

ASCE. (2013). *Pipelines for water conveyance and drainage*, ASCE, Reston, VA.

NSF (National Science Foundation)/ANSI (American National Standard Institute). (2016). "Drinking water system components—Health effects." *NSF/ANSI 61*, Ann Arbor, MI.

# CHAPTER 7
# DESIGN AND CONTRACT DOCUMENTS

## 7.1 GENERAL

This chapter addresses the design process for pilot tube and other guided boring projects, and the development of the contract documents, which will be used to complete the work. In most cases, the design and contract documents will be prepared by the owner's engineer. The design elements and process for the design of pilot tube and other guided boring projects are

- Design memoranda,
- Design calculations and analysis,
- Permit applications,
- Utility notifications/utility investigations, and
- Development of contract documents consisting of
  - General conditions,
  - Special or supplementary conditions,
  - Technical specifications,
  - Plans, and
  - Geotechnical investigation and reports.

The design is often completed in stages or phases that may include

- Feasibility evaluation,
- Preliminary design,
- Final design,
- Bid services, and
- Design support during construction.

Each of these design stages represents a general level of completion that is well established in the design community.

## 7.2 DESIGN MEMORANDA/TECHNICAL MEMORANDA

Design memoranda (DMs) and technical memoranda (TMs) may be prepared during preliminary design to explain

- Alternative alignments and construction methods considered;
- Hydraulic requirements;
- Potential environmental and permitting issues that must be addressed;
- Rationale for the design;
- Identified risks including settlement and jacking forces;
- Modeling and calculation results;
- Evaluation of pipe material options;
- Investigations of corrosion protection requirements; and
- Other issues specific to the project.

DMs and TMs help guide the final design effort. These memoranda are used to develop the contract plans and specifications provided to all bidders on the project. DMs and TMs are not typically distributed to the bidders.

## 7.3 CALCULATIONS

Once the pipeline's plan and profile have been established and laid out on the plans, calculations are needed to finalize the design. These calculations include the following:

- Size and slope of the carrier pipe;
- Distance between access point/shafts/manholes;
- Type and sizing of the shafts;
- Earth pressure diagrams and construction surcharge loads for shaft design;
- Internal supports, invert slab thickness, uplift, and wall penetrations of the shafts (see Chapter 5 for further discussion);
- Estimate of jacking forces and thrust resistance evaluation;
- Estimate of settlement;
- Earth loads on the pipe;
- In-use service loads; and
- Water discharge volume, if dewatering is used.

## 7.4 DESIGN CONSIDERATIONS

Whenever project requirements allow, the horizontal and vertical alignment should be selected, recognizing the limitations of the boring equipment and minimizing the risk associated with adverse subsurface conditions. For gravity-dependent installations, the engineer's ability to select the alignment may be restricted by the invert grade requirements.

In general, the pilot tube and other guided boring alignments should be selected to minimize the risk of encountering the following:

- Buried objects;
- Mixed-face conditions;
- Very soft/loose soils, lacking sufficient resistance required for steering, or the capacity to support the drill tools;
- Dense to very dense, hard, and/or cemented soils (unless use of a rotating cutting tool is possible);
- Saturated granular soils, unless groundwater control methods are employed; and
- Ground that may not provide sufficient long-term stability for the installation, when pipe settlement or flotation is possible.

The design must determine the size of the carrier pipe to be installed. For example, for a gravity system, the hydraulic calculations should incorporate flow monitoring data or hydraulic modeling data to size the pipe to pass the desired flows in combination with the slope of the pipe and the pipe's associated coefficient of friction. The slope of the pipeline must be considered and the line and grade tolerances of the proposed installation must be set.

The determination of the pipeline slope and distances between shafts will establish the pipe's profile and the depth of the proposed installation at the shaft locations. The pipeline alignment will need to be deep enough to avoid other buried utilities, provide sufficient depth to minimize settlement and heave of the ground during installation, and be located above or at the groundwater table, depending on the type of trenchless methods proposed.

It should be noted that most state and federal transportation agencies as well as railway owners have established specifications detailing pipe design criteria for trenchless crossings, which should be consulted during design, if applicable. These criteria often include details pertaining to minimum installation depth, length, and plan orientation of the installation.

### 7.4.1 Shaft Layout

Where shafts are needed, they should be shown on the plans. A detailed discussion of shafts and shaft design is provided in Chapter 5. The plans should include the following:

- Location and orientation of the shafts;
- Limits of the working space;
- Location of utilities, both overhead and underground; and
- Traffic control and access limitations.

See Sections 3.1.4 and 3.4.1 for further discussion on shaft planning and layout.

### 7.4.2 Estimate of Jacking Forces

Jacking forces must be estimated and evaluated based on anticipated ground conditions, assuming good workmanship and operational practices by the contractor to ensure the pipe and thrust reaction wall are designed to safely resist the jacking forces and that the jacking system has adequate capacity. Jacking forces should be evaluated using an industry-accepted approach during design. The specifications must require contractor submittals that confirm the jacking system has adequate capacity and that the pipe and thrust reaction wall has been designed to safely resist the anticipated jacking forces.

The contractor must provide pilot tube and boring equipment, pipes, gaskets, and connections that can resist the calculated jacking forces, and are consistent with the contractor's selected means and methods and anticipated ground conditions.

The criteria for determining jacking forces should take into account the ground conditions encountered in the geotechnical investigation and the baselines set in the geotechnical baseline report (GBR).

Jacking forces consist of two primary components: face pressure and frictional resistance. Methods to calculate jacking forces are provided in the following references:

- Bennett (1998);
- Bennett and Cording (2000);
- *Guide to best practice for the installation of pipe jacks and microtunnels* by the (UK) Pipe Jacking Association (1995);
- *Pipe Jacking and Microtunneling* by Thomson (1994);
- *Trenchless Technology for Installation of Cables and Pipelines* by Stein (2005); and
- *Trenchless Technology: Planning, Equipment and Methods* by Najafi (2013).

A variety of factors affect jacking forces, including:

- Soil type (clay, silt, sand, gravel);
- Soil consistency, which may be inferred from standard penetration test (N) values;
- Effective normal stress acting on thrust casing and pipe;
- Effective friction factor along the jacked pipeline;
- Lubrication;
- Overcut;
- Increased friction due to idle and downtime;
- Face pressures due to earth and hydrostatic pressures;
- Pipe surface roughness, dimensional consistency, surface porosity, and absorption; and
- Deflections at pipe joints.

Some of these factors are not explicitly taken into account when calculating jacking forces, using available methods. The key to reasonable accuracy is judgment in selection of appropriate ground properties based on a thorough geotechnical investigation and fundamental understanding of geotechnical characteristics that influence or control behavior.

When estimating jacking forces, operational factors can exert significant influence but are difficult to quantify. If lubrication is not used, or is ineffectively used, jacking forces can significantly increase. These and other operational factors are typically not within the engineer's control. Therefore, the engineer typically assumes proper equipment setup and good workmanship when making the jacking force calculations. A factor of safety is typically applied to account for these and other unknown factors. All assumptions should be clearly stated when making these and all design calculations.

If high jacking forces are encountered during pipe installation, pipe failure may occur. Any misalignment of the jacking pipe may reduce the ultimate load where pipe failure occurs because of point loads.

### 7.4.3 Estimate of Settlement

Settlement calculations should be undertaken and evaluated by the engineer. Existing utilities, roadways, railroads, foundations, structures, and other facilities must be protected against damage from ground movements resulting from shaft construction and pipe installation. The steps include:

- Identify facilities at risk for settlement.
- Establish recommended maximum allowable settlements for critical features through consultation with the facility owner.
- Evaluate potential settlements at each feature.
- Compare estimated settlements against recommended maximum allowable settlements for each feature and develop mitigation measures to reduce impacts to significant structures if necessary.
- Develop settlement instrumentation and monitoring plan for project, and ensure the plan is implemented during construction.
- Coordinate with permitting agencies and owners of critical facilities for successful implementation of a proactive program to avoid damage that may occur as a result of pipe installation and shaft construction.

The risk of ground movements (both vertical and horizontal) and the potential effect on the critical features should be evaluated. An acceptable approach is to identify sensitive and critical utilities, roadways, railroads, foundations, and other facilities that may be at risk from the pilot tube, boring, and shaft construction activities. Establish recommended maximum allowable settlements for critical facilities in coordination with facility owners. Suggested guideline settlement values for various facilities have been published (Wallin et al., 2008), but these values should be used as general guidance only, and refined through coordination with facility owners. If the evaluation results indicate high risks of damage to existing utilities or structures, it may be necessary to

- Relocate them;
- Protect them in place;
- Monitor movement and rehabilitate them if required after boring is complete;
- Adjust alignment of the new pipe installation; or
- Improve ground.

Ground movement associated with guided boring can be divided into two types: large ground movement at isolated locations and systematic settlements along the pipeline alignment. Large ground movements are typically the result of overexcavation at the face, which can occur because of improper setup of the boring equipment; rapid, unexpected changes in ground conditions, such as encountering loose sand or hard objects, mixed face, or artesian groundwater pressures; operator error; or some combination of these factors.

Systematic settlements are those associated with closure of the overcut and shield cut as the ground moves into the annulus. Additional systematic settlements associated with minor face losses may also occur. Systematic ground movement risks associated with boring should be evaluated, using methods described by Wallin et al. (2008). The approach approximates the settlement trough above a trenchless installation as an inverted normal probability distribution curve. Representative ground properties must be selected using judgment and the results of the geotechnical investigation. Systematic settlements are proportional to annular space and pipe outer diameter, and inversely proportional to ground stiffness, depth of earth cover, and offset distance from centerline of the bore. Calculated settlements increase as pipe diameter, shield cut, and overcut increase, and decrease as ground stiffness, depth, and offset distance increase.

Ground movement risk associated with shaft construction should be evaluated using approaches indicated in Clough and O'Rourke (1990). The results of the ground movement calculations, criteria for allowable limits, and the instrumentation and monitoring plan can be summarized in a TM during preliminary design.

Critical utilities and features should be monitored during construction using an appropriate instrumentation and monitoring plan. A formal instrumentation and monitoring plan may be appropriate for higher-risk projects. The plan should specify the critical features, locations and types of instruments, monitoring frequencies and durations, and maximum allowable and threshold values for settlements. The maximum allowable values should be less than values that could result in damage to the facility. Recent developments in technology have reduced the need for repeated personnel access to monitoring points, which may require lane closures or other disruptions. Examples include photogrammetric settlement surveys and laser profiling.

A preconstruction condition survey must be conducted before construction activities begin to establish conditions and protect against frivolous damage claims. Surveys should be continued during construction at established intervals, and for some time after construction is completed, until no additional movements are recorded between at least two successive surveys.

The contract documents should require that all settlement survey results be submitted to the engineer within a reasonably short time, such as 24 hours after the survey is conducted, so the results can be reviewed and any required actions implemented in a timely manner.

### 7.4.4 Earth Loads on the Pipe

Selection of appropriate pipe class or stiffness requires a reasonably conservative estimate of earth loads acting on the pipe. Computation of earth load is significantly different for pipe tunneled through undisturbed ground when compared with pipe placed within an open trench excavation. Pipe installed by boring a hole in the ground experiences lower earth loads than pipe installed in open trenches because of the ground's natural arching ability.

Earth loads on pipe installed by boring a hole in the ground are proportional to the diameter of the pipe and effective unit weight of the ground. Earth loads are inversely proportional to ground stiffness. For pipe installed in medium stiff to hard clay, or very dense sand, the earth loads acting on the pipe are typically less than the diameter of the pipe multiplied by the effective unit weight of the ground. For pipe installed in soft clay or loose to medium sand, the earth load acting on the pipe may approach two to three times the diameter of the pipe multiplied by the effective unit weight of the ground. Earth loads on pipe installed in stable rock and hard clay are negligible. Total overburden stresses should not be used to calculate earth loads on pipes except for relatively shallow installations in soft, squeezing clays, and loose to very loose sands. A conservative approach to determine earth load on a tunneled conduit is the conventional Marston formula when modified to include soil cohesion.

Refer to the following references for guidance on determination of appropriate design earth loads on pipe:

- ASCE (2007b)
- Bennett (1998)
- Bennett and Cording (2000)

Earth loads acting on pipe that protrudes into or across locations of planned or potential future cut-and-cover excavations (such as at jacking and receiving shafts, manhole locations, and service connections) are larger because of reduced arching action in cut-and-cover excavations. The more commonly used conventional Marston approach may be used to estimate earth loads acting on pipes in such locations. The axial jacking forces required for installation are typically the critical load case that controls pipe class or stiffness. The pipe manufacturer should be consulted regarding the earth, live, and installation loads that the pipe is capable of withstanding.

### 7.4.5 In-Use Service Loads

The pipes and joints, once installed, must be capable of safely withstanding, for the project design life, all in-use service loads. See Section 6.3.1 for a list of loads to consider.

### 7.4.6 Dewatering

The purpose of dewatering is to lower and control the groundwater to allow the contractor to perform the work in a safe manner, as previously discussed in Chapters 3 and 4. The owner's interest demands that the dewatering be done without delaying the schedule and without endangering personnel and equipment, that the methods do not impair the strength of the foundation soils, and that no damage to third parties results. Within these restrictions it is usually preferable to give the contractor maximum latitude to use ingenuity in reducing the cost of the work. The dewatering method is closely related to excavation operations, techniques of ground support, trenchless methods, and other factors. Unnecessary restrictions on the dewatering may escalate associated costs. In special circumstances, it may be advisable to specify certain dewatering methods.

The optimum form of dewatering specification will vary from one job situation to another. Dewatering specifications could include

- Specified results,
- Owner-designed dewatering systems, or
- Specified minimum systems.

The design of the dewatering system involves two important steps:

1. Evaluation of the magnitude of the dewatering project, including an estimate of the probable rate of inflow and power consumption.
2. Design of a system capable of providing the required groundwater lowering for the length of time needed for the construction that is to be carried out below the natural groundwater level.

For the design of any dewatering system, the geotechnical site conditions must be investigated and the data collected, as outlined in Section 4.2.2. The design of the dewatering system will be based on initial limited information and in accordance with good practices. More often than not, the actual dewatering system will be modified substantially from the original design, as the contractor adapts to the information developed during the installation of the system and follow-on observations and monitoring.

The design of a dewatering system will include a number of factors, such as

- Ground type, stratification, and thickness of the water-bearing formations;
- Permeability and transmissibility of the ground vertically and horizontally;
- Location and type of aquifers;
- Depth to which the water level must be lowered;
- Type of well system selected;
- Types and size of pumps selected;
- Effectiveness of the drawdown;
- Head loss in the piping system; and
- Location of the discharge point.

Estimates of the discharge volumes are typically needed during the design phase so discharge permits can be started and discharge points located. Also important during the design is to identify if there are contaminants in the ground or groundwater, and how the dewatering system will affect this contamination. If contaminated ground is present, prohibiting the use of a dewatering system should be considered.

## 7.5 CONTRACT DOCUMENTS

As with any civil works project, the bid/contract documents prepared by the owner and engineer should provide the contractor with the necessary information to prepare the bid, plan the work, and ultimately construct the project. These contract documents should include the following:

- General conditions,
- Special or supplementary conditions,
- Technical specifications,
- Plans, and
- Geotechnical information.

Regardless of whether a pipeline is to be installed by pilot tube and other guided boring methods, it is important that the contract documents be complete, clear, and concise, and prepared by individuals with experience in the specified methods and technology. Although there is much commonality between contract documents for pilot tube, guided boring methods, and any other pipeline projects, trenchless methods are specialty construction techniques; as such, some requirements unique to trenchless methods should be included in the contract documents.

The following sections discuss the preparation of the contract documents used for pilot tube and other guided boring projects.

### 7.5.1 General Conditions

The general conditions describe the contract conditions that govern construction projects for the owner agency or organization. The general conditions describe

- Work to be performed;
- List of bid items;
- Legal framework for enforcing the contract, including measurement and payment;
- Acceptance tests;
- Period of performance;
- Insurance requirements;
- Change order procedures;
- Dispute resolution; and
- Liquidated damages for failure to timely complete the project.

The general conditions should be reviewed for project-specific requirements and compatibility with the construction method. The general conditions of the contract specifications should include

- Precedents of documents;
- List of geotechnical information provided;
- Issue resolution processes;
- Schedule constraints;
- Local requirements; and
- A section permitting value engineering proposals for opportunities of significant economies and a better project.

Some of the preceding items listed may be found in the special or supplementary conditions.

### 7.5.2 Special or Supplementary Conditions

Special or supplementary conditions may be necessary for the particular project. These conditions may include the GBR and a differing site conditions (DSC) clause on some contracts.

### 7.5.3 Technical Specifications

The technical specifications include all the specifications that, taken together, establish the minimum requirements for the work to be performed, the materials to be used and installed, and the execution of the work. Technical specifications may be performance specifications or prescriptive specifications, sometimes referred to as means and methods specifications.

- Performance specifications establish the requirements for the product to be constructed, and leave means and methods to the contractor.
- Prescriptive specifications establish the requirements for the product to be constructed and the means and methods required to be used by the contractor to construct the specified product. The designer should be aware that prescriptive specifications create an implied warranty by the owner that if the contractor uses the means and methods required, with generally accepted industry good practices, the work product will meet the requirements of the contract.

Although not recommended, project technical specifications packages are sometimes a mixture of performance and prescriptive specifications, even within an individual technical specification, if the engineer and owner believe it is necessary to limit the range of contractor options. Strictly defined performance and prescriptive specifications are less commonly used.

The technical specifications package for a pilot tube and other guided boring methods project typically includes separate sections for each major item of work (e.g., the construction method, jacking pipe, shafts, geotechnical instrumentation and monitoring, contact/annular space grouting, groundwater control and disposal, installation of carrier pipe in casing if used, and backfill grouting between the casing and carrier pipe if used). The technical specifications should clearly describe each major item, relevant standards, design criteria, quality assurance and quality control (QA/QC), submittal requirements, work products, and execution.

The technical specifications for pilot tube and other guided boring methods projects should focus on performance rather than means and methods. As a general guideline, the technical specifications should convey the following information as a minimum:

- Minimum experience of contractor's key personnel and other requirements for the contractors;
- Jacking pipe materials, coatings, joints, and treatments acceptable to the owner;
- Capabilities and performance requirements of boring equipment;
- Constraints on construction sequence, noise, and work hours;
- Technical and contractual means of dealing with buried objects;
- Horizontal and vertical alignment tolerances;

- Shaft requirements;
- Contaminated ground and groundwater disposal requirements;
- Spoils disposal requirements;
- Settlement/heave criteria; and
- Required submittals.

In a design-bid-build contract, it is the engineer's responsibility to prepare a set of specifications to define the *Work* to be completed. The work of the project is documented as a series of specification sections. Each specification section defines the qualitative and administrative requirements for products, materials, quality, and workmanship for a specific element of the work on the project (ASCE, 2013). Each specification section is formatted into three parts. Part 1 is the general portion of the specification section, containing administrative and procedural requirements that only pertain to that section. Part 2 is the products portion, containing requirements for the materials and equipment for that particular section. Part 3 is the execution portion, containing requirements on the actual work elements of that section.

For consistency across the project specifications, the sections should be written in the same format. The specifications should be written and reviewed for consistency between sections. Requirements for specific work on the project should be established in one location. Common requirements for a separate piece of the work on the project could be included in a separate section, provided there is cross-referencing in both sections. Cross-referencing is important to tie requirements together between specification sections.

### 7.5.4 Plans

The design plans must clearly and concisely show the work to be done. As a general guideline, the plan and profile drawings should convey the following information as a minimum:

- Number of drives, lengths, diameters, slopes, inlet and outlet elevations at manholes, and shaft locations;
- All known aboveground and below-ground utilities and structures, both active and abandoned;
- Subsurface exploration locations and conditions;
- Shaft locations and work areas;
- Geotechnical instrumentation locations;
- Locations of known soil and rock horizons or contamination requiring special treatment or disposal requirements; and
- Groundwater depth and location.

The plan and profile sheets should show the geotechnical boring logs. By doing so, the geotechnical information and actual plan and profile are to the same scale and show potential challenges along the drives. In addition to plan and profiles, as many contract plans as necessary should be prepared to convey the following information:

- Survey control points for project layout;
- Limits of work areas at all shaft locations;
- Ground improvement details;
- Additional work and staging areas;
- Rights-of-way and/or easements;
- Geotechnical instrumentation details and schedules with tabular locations;
- Minimum requirements and design criteria for temporary shafts;
- Pipe details;
- Traffic (vehicular and pedestrian) management plans;
- Work necessary for protection of adjacent structures;
- Utility relocations;
- Standard details per the appropriate agency (e.g., manholes, manhole lids, driveways, curb, and gutters); and
- Erosion control measures and details.

Appropriate notes should advise the contractor of the need to require utility locates, confirm all crossing and adjacent utilities within the tunnel zone of influence, and protect all utilities from damage. Additional notes should convey information necessary to guide the contractor in the execution of its work.

Protection of critical adjacent structures and utilities should not be left entirely to the contractor. Where special underpinning, grouting, or other measures are designed by the engineer, sufficient detail should be given so that the contractor can bid and construct the designed system.

### 7.5.5 Geotechnical Investigations and Data

Preliminary and detailed geotechnical investigation and reporting requirements are described in Chapter 4. All geotechnical data collected during the various phases of the project should be compiled and provided to all bidders on the project. Some agencies will compile and provide geotechnical data in an electronic format.

The owner of the project might request the engineer of record to write a GBR, depending on the level of complexity and risk. This report or letter should follow the table of contents outlined in *Geotechnical Baseline Reports for Construction: Suggested Guidelines* (ASCE 2007a). All parties to the design including the owner should participate in the preparation of the GBR. The GBR must be provided to all bidders on the project.

## 7.6 DISPUTE RESOLUTION

A method of resolving disputes between the owner and contractor should be included in the contract documents. Even with the best engineering effort and exploration program, trenchless work, whether conventional or small-diameter tunneling, is often fraught with unforeseen underground conditions. Many owners and contractors do not fully understand the unpredictable nature of subsurface conditions and the impossibility of accounting for all of the conditions possibly encountered. It is imperative that any underground contract contain a DSC clause to allow the contractor to be compensated for extra costs involved without being forced into a breach of contract. This is required of federal contracts and should also be included for nonfederal government work. Typically, the federal clause should be used because it has been tested in court and found to be fair and enforceable.

Adversarial relationships should be avoided whenever possible. It is in everyone's best interest to resolve conflicts quickly, fairly, and equitably. The contract documents should provide for mediation or arbitration before litigation. The following are generally considered positive steps:

*Clear, unambiguous contract documents.* Owners should realize that spending a little extra time and money during the design phase for complete explorations and design tends to pay large dividends in decreased change orders and extra costs during construction.

*DSC clause and GBR.* To help clarify exactly what constitutes a DSC on a particular job, trenchless designers should adopt the GBR. This document is a part of the contract documents and establishes baseline values for conditions to be expected during construction, such as frequency of encountering cobbles and boulders, or groundwater inflows to be expected during construction. The joint ASCE/AIME publication *Geotechnical Baseline Reports for Construction: Suggested Guidelines* (ASCE 2007a) describes GBRs in great detail. Readers are strongly urged to refer to this document for more information on GBR history, rationale, and preparation. Please note that the guidelines are written for large-diameter tunnels and some provisions may need to be modified for small-diameter tunnels.

*Fair and timely dispute resolution plan.* The time for response between engineer and contractor on requests for information and notices of dispute should be specified so turnaround is kept to a practical minimum. The quick resolution of disputes may reduce or eliminate the schedule impacts of a long-drawn-out or untimely, nonresponsive process.

*Alternative Dispute Resolution (ADR).* Once the preliminary dispute resolution measures have contractually been exhausted, contracts should allow for some ADR procedure prior to the long and costly litigation process. Many in the tunneling industry [such as ASCE, the American

Arbitration Association (AAA), the Underground Technology Research Council, and others] have developed ADR techniques that can be applied to the trenchless industry (AAA 2015). Such elements as dispute review boards (DRBs) and escrow bid documents might be considered depending on the project's cost, time frame, and complexity. The reader is directed to the reference published by Underground Technology Research Council (1991), *Avoiding and Resolving Disputes during Construction*. This document, the forerunner of the previously described ASCE/AIME 2007 publication, discusses GBRs (then called GDSRs), escrow bid documents, and DRBs. The use of DRBs has been further refined and explained since the 1991 publication. The reader is directed to the Dispute Resolution Board Foundation website at www.drb.org.

The daily cost of construction is expensive for the contractor and the owner. Use of a DRB can reduce the cost of nonproductive work and the effect of disputes on the work progress. When a dispute delays or stops work, the cost to all parties increases. Use of a DRB is a good method to help resolve a dispute quickly and keep the project moving forward. The benefit of the DRB is that it brings in objective individuals familiar with trenchless and shaft construction methods and the project itself to provide input and expert opinion of causes of the dispute. It is also recommended that the DRB meet on a regular basis (a minimum of quarterly because of the typically short duration of a trenchless project) to remain informed about the project status, even if there are no disputes; knowing the DRB is current with the project conditions helps the owner and contractor keep an open conversation about the project.

## 7.7 CONTRACTOR QUALIFICATIONS

Prequalifying a contractor and/or subcontractor to conduct pilot tube and other guided boring work should be considered for projects with identified high-risk elements (see Section 3.5.4). It is acknowledged that technical prequalification is not always necessary. In addition, prequalification may not be possible because of public bidding laws. When prequalification is not feasible, proof of minimum experience requirements for project manager, operators, and superintendents should still be stipulated as required submittals with the bid or in the contract documents.

Although the experience of the design team and a good set of contract documents are important to the success of a pilot tube and other guided boring project, the experience of the contractor is critical. It is desirable to prequalify specialty contractors or subcontractors based on technical/experience criteria. The goal is to ensure that the work would be done by a contractor's or subcontractor's key on-site personnel, who are familiar with the specific challenges and potential problems inherent with the pilot tube and other guided boring methods. In some instances, the contractor might supplement its experience through use of qualified technicians provided by the equipment manufacturer to train the contractor's operator and crew by initially operating the equipment.

The specific mechanism for prequalification must be established on a case-by-case basis, depending on the risks associated with the project. The following criteria might be considered for prequalifying a contractor or subcontractor performing the work:

- Proof of bonding capability;
- Minimum of five successful drives involving pilot tube and other guided boring methods by the contractor in similar ground conditions;
- Minimum of 1,000 ft (300 m) or five drives for personnel operating the boring equipment with similar equipment to that proposed for the particular work; and
- Safety program and performance.

Some owners and engineers consider the prequalification of a bidder and its workforce (for pilot tube and other guided boring, shaft design and construction, and associated ground improvement work) a prerequisite for the owner to open the bidder's bid, for one or a combination of the following reasons:

- It is unfair for inexperienced contractors to learn or experiment with public funds, particularly on specialized or difficult work.
- Underground construction work is inherently risky and can be dangerous, particularly for contractors and a workforce with insufficient experience.
- Past project delays, damage to public and private properties, and exposure to third-party liabilities because of ground subsidence related to tunneling errors, problematic shaft design and construction, inadequate equipment ingress and egress, inadequate shaft perimeter support, unsafe and messy work sites.
- Past project problems related to estimator errors during bidding, including inadequate shaft sizes, and incomplete costing of shaft support.
- Past construction problems caused by contractor using inexperienced personnel to lay out shaft locations, resulting in ingress problems.
- Substantially higher inspection and construction management costs associated with inexperienced contractor or workforce.
- With experienced and qualified contractors and workforce, many projects have been successfully completed in difficult ground and groundwater conditions.

An example of a bidder's prequalification process is as follows:

- Owner requires that all contractors requested to, and interested in, submitting a bid for this project submit the qualification statement by the date specified in the Notice to Contractors.
- Prospective bidders must provide all requested information and answer all questions fully and explicitly.
- Only bids from qualified contractors having all the required prequalified workforce members and track records, or contractors with prequalified subcontractors shall submit a bid. Bids submitted from contractors that have not been prequalified will be returned.
- Each prospective bidder submitting a Bidder's Statement of Qualification will be notified by the owner either by mail or fax, prior to bid opening, indicating which individuals are deemed qualified and unqualified. The bidding contractor may then select, from those individuals deemed qualified by the project owner, to assemble its workforce.
- General contractors cannot use previous subcontractors' experiences on past projects to qualify themselves individually. Only self-performed work experience obtained prior to the Bidder's Statement of Qualification deadline date may be used as part of an individual's qualifying work experience. The owner, at its discretion, may ask for clarification or additional supporting data after this deadline.
- If the general contractor does not have the qualifications to self-perform, the general contractor must designate a qualified subcontractor it intends to use in its bid submittal.
- The final decision to accept or reject a contractor, a contractor's workforce member, or subcontractor based on the information provided in this Bidder's Statement of Qualification lies solely with the owner.
- Requirements of this Bidder's Statement of Qualification remain in force through project completion.

## REFERENCES

AAA (American Arbitration Association). (2015). *The AAA guide to drafting alternative dispute resolution clauses for construction contracts*, New York, 29.

ASCE. (2007a). *Geotechnical baseline reports for underground construction: Suggested guidelines*, Reston, VA.

ASCE. (2007b). *Gravity sanitary sewer design and construction*, 2nd Ed., Reston, VA.

ASCE. (2013). *Preparation of construction specifications for civil projects*, Reston, VA.

Bennett, D., and Cording, E. (2000). "Jacking loads and ground deformations associated with microtunneling." *Proc., North American No-Dig*, Anaheim, CA, NASTT, Cleveland, 17–36.

Bennett, R. D. (1998). "Jacking loads and ground deformations associated with microtunneling." Ph.D. dissertation, Univ. of Illinois, Urbana-Champaign, IL, 486.

Clough, G. W., and O'Rourke, T. D. (1990). "Construction induced movements of insitu walls." *Proc., Design and Performance of Earth Retaining Structures*, ASCE, Cornell Univ., Ithaca, NY, 439–469.

Najafi, M. (2013). *Trenchless technology: Planning, equipment and methods*, McGraw Hill, New York.

Pipe Jacking Association. (1995). *Guide to best practice for the installation of pipe jacks and microtunnels*, London.

Stein, D. (2005). *Trenchless technology for installation of cables and pipelines*, Stein & Partner, Berlin.

Thomson, J. (1994). *Pipejacking and microtunneling*, Chapman & Hall, New York.

Underground Technology Research Council. (1991). *Avoiding and resolving disputes in underground construction*, 2nd Ed., ASCE, New York.

Wallin, M., Wallin, K., and Bennett, D. (2008). "Analysis and mitigation of settlement risks in new trenchless installations." *Proc., North American No-Dig*, Dallas, TX, NASTT, Cleveland.

# CHAPTER 8
# CONSTRUCTION

## 8.1 GENERAL

The following sections are based on the traditional design-bid-build process. It should be noted that other project delivery methods are available. Once final design is approved, funding is in place, and the permits and easements acquired prior to the start of construction are known, the project owner will put the project out for bid.

## 8.2 BIDDING

On any given project, a general contractor oversees all elements of the work. The trenchless and shaft portions of the project may be smaller elements of a larger project. Pilot tube and other guided boring installations involve specialty work and often are performed by a subcontractor. Shaft construction and dewatering may also be performed by a specialty subcontractor. If the work is not self-performed, the general contractor needs to coordinate the various subcontractors to ensure project success.

During bidding, interested bidders (general contractors and subcontractors) should review the bid documents, which may include the following:

- General conditions;
- Special or supplementary conditions;
- Technical specifications;
- Plans; and
- Geotechnical information, including subsurface exploration logs and geotechnical reports.

It is recommended that the project owner schedule a prebid meeting with a site visit and provide an opportunity for all interested bidders to attend. This meeting provides an opportunity for bidders to evaluate the existing site conditions, including possible utility interference, traffic control challenges, site access, adjacent construction activities, and other site constraints. For projects that are considered difficult, complex, or have important constraints/requirements to be known or addressed by the bidders, the owner should consider making the prebid meeting and site visit mandatory.

A thorough understanding of the subsurface conditions allows the contractor to select the appropriate construction means and methods. This dictates the success or failure of the installation. Therefore, a key consideration during the bidding phase is how the contract language presents and characterizes the geotechnical information. The contractor should also determine whether the contract documents include a differing site condition (DSC) clause. These items greatly influence the contractor's risk and, therefore, greatly influence the contractor's bid price.

After reviewing the contract bid documents, the general contractor will prepare and submit a bid based on the estimated cost and schedule to build the project. Responsive bidders will include in their bid the cost for any subcontracted work to be performed by others. Each contractor's bid will include appropriate markup for risk, overhead, and profit.

## 8.3 SUBMITTALS

The contractor performing the pilot tube or guided boring work will prepare and provide detailed submittals in accordance with the requirements of the specifications. The submittals may include working and shop drawings, catalog cuts, calculations, schedules, and written work plans. The extent of the submittals depends on the complexity and specific risks associated with the project. The timing of the submittals varies and includes preconstruction, construction, and postconstruction documents.

The pilot tube or guided boring submittals might include the following:

A. Preconstruction Submittals
- Qualifications for the following personnel
  - Trenchless key personnel
  - Professional engineer experienced in shaft design
  - Surveyor familiar with working underground and in shafts
- Pilot tube and guided boring equipment
  - Method statement (general description of how the system works and operates)
  - Jacking system
  - Guidance and steering system

- Construction Work Plan
  - A narrative describing the contractor's approach to the work. It often includes a sequence of planned construction, a schedule and a discussion of the individual drives and the geotechnical conditions expected to be encountered during each.
  - The work plan may also include a geotechnical instrumentation and monitoring program as well as the contractor's plans for mitigation of potential issues such as surface settlement, heave, encountering obstructions, or a stuck pipe.
- Jacking pipe details
  - The diameter, strength, wall thickness, joint design, and other fabrication details of the jacking pipe.
- Site layout plan
  - A site layout plan ensures that all aspects of the work remain within the established easements and rights-of-way. In addition, it can provide a check of proximity of the equipment to nearby utilities (underground and overhead), roadways, or railroad tracks to ensure a safe operation.
- Lubrication system details
  - As in any pipe jacking operation, it is important to lubricate the ground/pipe interface to reduce the jacking force required to complete the work. Details should include how the contractor plans to lubricate the pipe string, including the products it intends to use.
- Spoils management plan
  - Consideration should be given to any special material handling and disposal requirements, as well as documentation of the final disposition of the excavated material. The contractor should provide details of the spoils disposal location, and transportation methods.
- Estimate of jacking force
  - The estimated jacking force is used to confirm that the jacking equipment is adequate, that the jacking pipe can withstand the axial load that will be applied during installation, and that the excavation support can withstand the forces required to enable the installation.
- Shaft design
  - The shaft should be constructed in accordance with Occupational Safety and Health Administration (OSHA) requirements, including the provision of a professionally engineered design when required.
  - If necessary, the jacking shaft design should include provisions for a reaction block (or similar) intended to resist the loads applied by the jacking equipment.

  - The shaft design should be developed to accommodate the penetrations required to launch and receive the pilot tube and guided boring methods tools, as applicable.
- Dewatering plan
  - The pilot tube and other guided boring methods require dry shafts and an alignment at or above the groundwater table. Consideration must be given as to whether it is permissible to lower the water table to achieve that result. It is important that the dewatering effort reduce the groundwater levels assumed by the shaft designer.
  - The dewatering plan should provide a summary of dewatering methods, estimated pumping volumes, groundwater treatment and disposal methods, and an estimate of the time required to lower the groundwater table to the desired elevation.
- Contingency plan
  - Trenchless pipeline projects involve unique contingency planning requirements. The contract documents generally require the contractor to submit contingency plans for specific events that could occur. The contingency plan explains how the contractor will react to specific events and how it will mitigate the risk associated with this event. Some of these specific events could include:
    - Excessive ground movement (subsidence and heave);
    - Excessive deviation from the target path;
    - Damage to existing utilities;
    - Dewatering problems in the jacking or receiving shaft; and
    - Encountering obstructions.
- Safety plan
  - All work must be developed in accordance with OSHA standards, or more stringent local/regulatory standards when applicable.
  - In some cases, depending on the diameter of the pipeline installation, internal access to the casing by construction personnel may be advantageous. It should be noted that this might require development of a site-specific confined space plan in accordance with the OSHA "Construction Standards for Excavations," 29 CFR Part 1926.
- Ground stabilization details (if required)
- Preconstruction surveys
  - One method of risk management is to conduct a preconstruction survey of surface features such as sidewalks, pavement, and curb lines as well as facades of nearby buildings and other structures to document the existing preconstruction condition of nearby structures above, at, and below grade. This may also include the interior lower floors of nearby sensitive structures.

  - The preconstruction survey could include video recording with narration, photographic evidence of conditions, and possibly geodetic survey of specific features.
  - Preconstruction surveys are intended to protect all parties from third-party claims.

B. Construction Submittals
- Construction records provide details of the work performed during construction to allow the owner to determine whether the pipe installation is progressing satisfactorily, and if not, whether changes to the methodology can be implemented. These records could include:
  - Construction progress, including documentation of any unusual occurrences or construction challenges encountered;
  - Photographs;
  - Schedule updates;
  - Survey measurements and interpretation;
  - Geotechnical instrumentation readings and interpretations;
  - Volume of lubrication used and when it was used;
  - Geotechnical conditions encountered;
  - Dewatering details;
  - Jacking loads experienced; and
  - Volume of material excavated.

C. Postconstruction Submittals
- Record drawing of as-built pipeline
- Ground movement data and summary table
- Postconstruction surveys
  - These surveys may be required to document the postconstruction condition of nearby structures above, at, and below grade, to determine whether any change in condition has occurred during the construction activities.
  - The postconstruction survey should be compared with the preconstruction survey to document any changes that may have occurred during construction operations.

## 8.4 MEASUREMENT AND PAYMENT

Owners may request that the work be performed on a lump sum cost basis, or as a unit cost basis. A lump sum basis means the owner will pay one fixed amount for the anticipated work. The number of lump sum items can vary on a project. The owner establishes the limits of the work included under each lump sum item. It should be noted that the owner might request a single lump sum cost for the entire project, or a series of individual lump sum costs. Lump sum prices can be further defined with partial payments based on a *Schedule of Values* associated with each lump sum.

A unit cost basis means the owner will pay the contractor on a per-unit basis, using rates established within the contract. For unit price bids, the owner establishes a bid schedule to obtain pricing for all work associated with the pilot tube and other guided boring methods installation. This bid schedule provides the bidders the opportunity to establish their lump sums (schedule of values) or unit prices for individual work items. A typical unit cost basis bid schedule could include

- Mobilization (generally lump sum, covering both mobilization and demobilization);
- Jacking and receiving shafts (lump sum, per vertical foot, or per each);
- Dewatering (lump sum);
- Installation of casing, if required (per linear foot);
- Installation of carrier pipe within casing, if required (per linear foot);
- Direct jack installation of carrier pipe, if required (per linear foot);
- Use of pilot tube for probing, if required (per linear foot);
- Rescue shafts (lump sum or per vertical foot);
- Manhole/access installations (per each); and
- Site restoration (lump sum).

Mobilization expenses should include transportation costs of equipment, temporary facilities as required by the contract documents, and setup of the work sites. A separate bid item for mobilization should be included because the contractor tends to have many up-front costs before actually installing pipe in the ground. These front-end costs typically include permit applications required by the contract documents, preparation and submission of technical submittals, and preconstruction and postconstruction surveys. Mobilization is normally a lump sum bid item or a percentage of the contract amount and typically also covers costs associated with demobilization. If, as a result of a changed condition, the contractor incurs additional mobilization costs, these costs should be paid in addition to the original mobilization bid item.

Jacking and receiving shaft costs should include design of the shafts and preparation/submittal of shop drawings, as well as the subsequent construction of the shafts in accordance with the approved design and in accordance with the contract documents. Separate bid items for jacking and receiving shafts are warranted because the work could be completed by the general contractor or by the subcontractor. In addition, the depths of the shafts may vary, so some projects include a bid item for each shaft as a lump sum. Typically, items of work required to prepare the shaft for the boring work such as installation of the entry/exit seals, invert leveling slab, and setting of guide rails are paid under the pilot tube bid item.

Pilot tube and other guided boring methods can be used to guide the installation of casings or the direct installation of the carrier pipe. Other applications exist as well, such as probing the geology in advance of other trenchless work or for the installation of grout injection tubes, or installation of horizontal dewatering wells. Each of these scenarios has a different typical payment method.

If a permanent casing is required, there is usually one pay item for the casing and another for the carrier pipe/conduit to be installed within the casing. A separate bid item should be considered to pay for pipe material upon delivery because the pipe is typically specially made for the contractor's equipment and delivered well in advance of installation.

Casing is usually measured and paid for by the distance from shaft wall to shaft wall, whereas the carrier pipe inserted within the casing is usually measured from center of manhole to center of manhole. Likewise, a direct jacked carrier pipe is usually paid from center of manhole to center of manhole. If a pilot tube is installed for guidance, it is considered incidental to the overall pipeline installation. If the contractor has purchased additional sections of pipe that are not installed, it is considered incidental to the other pay items and is not reimbursed separately by the owner.

When the pilot tube is used for probing, it is usually paid for by the linear foot of pilot tube installed. When pilot tube is used for the installation of grout tubes, it is usually considered to be incidental to the grout tube installation pay item and is not paid for separately. The same applies for use of pilot tube to install horizontal dewatering wells.

Provisions should be established in the contract to cover the cost and time associated with rescue shafts. A rescue shaft may be needed for obstruction removal, DSCs, equipment failure, or excessive deviation from the design line and grade. The need and value of a rescue shaft should be defined based on the anticipated site and ground conditions. Payment for a rescue shaft may be an actual contractor bid item that provides for payment per vertical foot of shaft excavation, or per shaft. Alternatively, rescue shafts can be paid as time and materials under a fixed price allowance for all bidders. The amount of the allowance should reflect the project risk. For most projects, the location and level of effort required to build a rescue shaft are unknown. An allowance item paid by time and materials apportions risk fairly. Compensation for a rescue shaft should be based on the evaluation of cause. Compensation should include the contractor's downtime costs and ownership/leasing costs of the guided boring equipment.

Separate bid items should be established for the installation of manholes and other access facilities in any of the shafts constructed for the trenchless work. This work is typically completed by a different subcontractor than the pilot tube and guided boring contractor.

Site restoration expenses should include repaving the road surface, sidewalks, fencing, and landscaping at the shaft locations, and bringing the work spaces back to their original conditions, generally in accordance with local jurisdiction requirements.

Demobilization expenses should include removal of equipment, transportation costs of equipment off-site, decommissioning of temporary facilities, and cleaning work sites. In most cases, the cost of demobilization is covered under the mobilization bid item.

If site conditions are different from those described in the contract documents and the contractor can show the different conditions adversely affected the work, the contractor will be entitled to some form of compensation. This is described further in Section 8.9.

## 8.5 JOBSITE LAYOUT AND EQUIPMENT SETUP

Major permits, easements, and rights-of-ways are procured in advance of construction, usually during the design phase of the project. The contractor may be required to obtain permits such as street occupancy, spoils discharge, or dewatering discharge based on how the contractor plans to execute the work.

All overhead and underground utilities that may affect construction should be located and shown on the plans by the engineer as part of the contract documents. If the contractor will be expected to locate utilities through specialized survey, or *soft-dig* techniques, this should be noted in the contract documents and included as a separate bid item. Prior to digging the shafts and commencing boring operations, the contractor should call the local one-call center to survey and mark all underground utilities along the project alignment. The contractor should verify the exact location and depth of the utilities via careful test pit excavations at those locations where there are apparent utility conflicts with the project work. Test pits at shaft locations should encompass the full perimeter of the shaft footprint. Manual excavation may be required for the last few inches from the existing utilities to avoid damaging the utility. Vacuum excavation is an excellent option to expose utilities with little surface disturbance and minimal risk to the existing utility.

Those utilities that will affect construction and cannot be protected or supported in place within any shaft or pipeline alignment should be coordinated in advance with the utility owner to be temporarily or permanently relocated.

Sufficient space is needed around the jacking shaft for the following:

- Initial space for shaft construction equipment and materials;
- Pilot tube and other guided boring methods equipment and tooling;
- Lubrication mixing and pumping system;
- Power supply generators;

- Storage and layout of the jacking pipe or pullback pipe;
- Support equipment such as backhoe, loader, or crane;
- Temporary stockpiling, loading, and hauling of excavated material; and
- Additional space for typical construction operations, including dewatering.

Refer to Section 3.1.4 for recommended work space requirements. The work space includes both the space available at grade and the space inside the shaft. The work space at grade should be of sufficient size for the construction of the shafts, the safe operation of all equipment, and for storing the materials necessary to complete the installation. The shafts should be of sufficient size for the free and safe movement of personnel and equipment during the setup of the equipment as well as during the execution of all the steps of the pilot tube and other guided boring methods process, including the movement of materials and equipment into the shaft and the removal of spoils from the shaft.

The layout of the equipment and materials around the shaft should be planned for the efficient and safe operation of the equipment and the movement of the material with minimum handling.

The work space should be limited to the easements and/or rights-of-way unless prior written consent is obtained from an adjacent property owner, usually after proper compensation is made.

Local authorities (e.g., emergency services, transit, and other agencies) should be notified of upcoming work activities in advance of mobilizing to the work site. Pedestrian, vehicular, and emergency traffic considerations, noise restrictions, and the needs of the local residents are just a few of the items that may require attention. Refer to Section 3.2.6 for further discussion of special considerations to be identified and addressed.

Figs. 8-1 through 8-4 provide examples of typical jobsite layouts. The figures show isolated sites with no pedestrian, vehicular, or emergency traffic.

## 8.6 SURVEY

Prior to start of the work, the contractor will require access to established project survey benchmarks. In addition, it is helpful to provide the contractor with electronic copies of the project survey files. These will help ensure the installation is completed at the correct location, and at the required project line and grade.

As described in Chapter 2, the pilot tube and other guided boring methods guidance systems use a digital display (e.g., tablet, computer), a

*Fig. 8-1. Typical pilot tube jobsite work space along a major highway to install a sanitary sewer crossing under the highway (photo courtesy of National Clay Pipe Institute)*

*Fig. 8-2. Typical jobsite work space along a street with a crane, shaft, equipment container, generator, and lubricant mixing tank (photo courtesy of Northeast Remsco Construction, Inc.)*

theodolite, and an illuminated target to provide information to the boring machine operator as to the actual location of the pilot tube head compared with the desired line and grade. Figs. 8-5 and 8-6 provide examples of guidance system components. Using that information, the operator rotates the head as the tube is advanced to steer the pilot and maintain the drive within the line and grade installation tolerances.

*Fig. 8-3. Typical jobsite work area with space for the storage of the carrier pipe (photo courtesy of Northeast Remsco Construction, Inc.)*

*Fig. 8-4. Typical jobsite workspace with a gantry crane to lower the jacking pipe into the jacking shaft and the pipe with staged hoses to reduce makeup time (photo courtesy of National Clay Pipe Institute)*

Vertical accuracy relies on having a properly calibrated theodolite fastened to a secure stand at a specific elevation. After which, the desired slope is *dialed in* to the theodolite.

Horizontal accuracy depends on precise survey marks placed at the front and the rear of the launch shaft. The horizontal alignment of the theodolite and the jacking frame are established using plumb bobs suspended from a string pulled taut between the survey marks.

*Fig. 8-5. Camera-mounted theodolite used as part of the guidance system and setup in the jacking shaft on an independent stand isolated from the jacking frame (photo courtesy of Northeast Remsco Construction, Inc.)*

*Fig. 8-6. Digital display with the image from the camera-mounted theodolite looking at the light source at the steering head of the pilot tube (photo courtesy of Northeast Remsco Construction, Inc.)*

A mistake in elevation at the jacking shaft will be carried along the entire length of the drive. However, a small error in horizontal layout at the jacking shaft will be multiplied many times over the length of the drive. Accordingly, pilot tube and other guided boring methods require accurate site surface survey information to be successful. The location information used by the operator is only as good as the layout information provided by the surveyor and the care with which the crew transfers that information into the shaft.

## 8.7 JACKING AND RECEIVING SHAFTS

As described in previous chapters, a pilot tube and other guide boring operation requires two shafts: a jacking shaft and a receiving shaft. The shafts should be located at a safe distance from existing structures and utilities to avoid property damage and reduce risk of injury.

The locations of the jacking and receiving shafts should be carefully selected, usually by the design engineer. Ideally, the shafts should be planned at manholes or lateral connections in sewer lines and at fire hydrants or gate valves in water applications or at splice points for high-voltage cables. Also the distance between the jacking shaft and receiving shaft should be minimized whenever possible so that friction forces do not prevent the successful installation of the pilot tubes or thrust casings, or exceed the jacking capacity of the boring equipment or the jacking pipe.

In most cases, selection of the shaft support will be at the discretion of the contractor, subject to performance criteria established in the project specifications. The shafts may need to have an engineered earth support system if space is limited or the shafts are deep. In such cases, the engineered support system should be specific to the site, prepared by an experienced engineer, and submitted for review by the designer. However, if sufficient room is available, sloping of the shaft walls may be sufficient. Regardless of whether the shaft has an earth support system, all elements of shaft design and construction must comply with OSHA standards. Available shoring methods may include: trench boxes, steel frames and plates, and corrugated pipes. Engineered earth support systems may consist of soldier piles and lagging, precast concrete rings, precast concrete caisson, and steel sheet piles.

In some cases, it is feasible to design the final maintenance access points (e.g., manholes) to serve as the launch or receiving shaft. Refer to Chapter 5 for further discussion on the use of temporary and permanent shafts.

The jacking shaft should be long enough to accommodate the pilot tube and guided boring jacking frame and extensions, the thrust block, and the length of the jacking pipe, pilot tube, or thrust casing. It also should be wide enough to accommodate the boring equipment operator plus workers for adding pilot tubes, thrust casings, and jacking pipes as well as for removing spoils.

After underground and overhead interference are removed or otherwise protected, and the location of the installation is confirmed with the site survey, shaft construction can begin. Once the excavation has been completed, a stable subgrade and working surface is needed in the shaft bottom. This stable surface can be accomplished by using crushed stone, gravel, steel plates, wood beams, concrete floor, or combination as geotechnical considerations warrant for the project. The floor of the shaft is set to the proposed slope of the bore unless the jacking equipment has grade adjustment capability.

The effect of the groundwater elevation on shaft design and construction should be considered. Dewatering efforts may be required to lower the groundwater at the shafts as well as along the alignment. Refer to Chapter 4 for further discussion on groundwater. The contractor must be prepared to accommodate the dewatering equipment during pilot tube or guided boring operations and shaft construction. As is the case with every dewatering system, the contractor should take measures to prevent migration of ground. During dewatering operations, treatment and discharge plans should be developed that comply with regulatory agency requirements.

Surface water should be managed so that the shaft subgrade remains dry and stable. This can be accomplished using ditches and/or berms to divert the overland flow around the shaft. A sump pump may be needed to handle rainwater in the shaft.

During pilot tube advance or other guided boring operations, the jacking thrust is usually transferred to the back of the shaft, where it is resisted by the passive pressure of the soil behind the shaft. A supplemental backing plate or thrust (reaction) block may be installed against the rear wall of the shaft to help distribute the thrust force exerted by the boring machine to prevent shaft damage. The backing plate may be steel plate, timber, concrete thrust block, or a combination thereof. The thrust force exerted by the boring machine distributed through the backing plate and thrust block (if used) must not exceed the passive earth pressure of the soil immediately behind the shaft with an acceptable factor of safety. Adequate care should be taken to ensure that the thrust pressure developed by the operation does not affect existing utilities near the bore shaft, or adjacent surface grades.

Where insufficient passive pressure exists to withstand the jacking thrust, other means of thrust restraint must be employed, such as: ground improvement, steel shanks driven into the soil, bolts anchored into the concrete floor, driven steel piles, or a wall constructed integral with a concrete floor extending below grade.

## 8.8 SAFETY ISSUES

It is the responsibility of the contractor to maintain a safe working environment in and around the shaft and to comply with the requirements of OSHA 29 CFR part 1926, and local and state agencies having jurisdiction. There are specific requirements for shaft construction, including personnel protection using surface barricades, traffic management to protect workers within the work zone, shaft access and egress, and ventilation in the shaft for personnel. The standard safety procedures adhered to in typical open-cut construction should also be followed on trenchless projects.

Manufacturer's safety and operations manuals should also be consulted for specific safety requirements unique to the equipment provided. Workers should understand the components of the pilot tube and guided boring equipment, with special attention to the moving parts in the system. It is critical to ensure that all personnel are trained to safely perform the boring operations. Workers should also be trained in the proper use of personal protective equipment.

For larger-diameter casings, internal access to the casing by construction personnel may be necessary to facilitate installation and a determination whether a confined space exists. If so, all personnel accessing the casing should be suitably trained for confined-space work. Further, adequate lighting and ventilation should be provided.

## 8.9 DIFFERING SITE CONDITIONS

The extent to which the subsurface conditions, including geotechnical conditions and potential buried objects (see Section 3.2.4), are identified prior to construction depends on how comprehensive the site investigation is during the design phase of the project. If they are identified in advance, a plan to overcome the challenges associated with the geotechnical conditions or buried objects can be developed during the design and construction phases. Unfortunately, not all geotechnical conditions or buried objects are identified prior to construction. This might force the pipeline to deviate from its intended alignment or stop the pilot tube or other guided boring methods process altogether.

Obstructions are defined as any buried object that lies completely or partly within the cross section of the boring equipment and that impedes continued forward progress of the pilot tube or other guided bore along the design path within allowable tolerances.

The contract documents should allow for adjustments in the project duration and price if unforeseen conditions are encountered. The contractor should be required to explain how the condition encountered differed from those documented in the contract. The contractor should document the time and cost impacts to the pilot tube and other guided boring methods operation. Timely notice to the project owner is important to successfully evaluate the differing conditions and to determine the appropriate course of action, as applicable. It is typical that the contractor not proceed with any extra work associated with the differing condition until instructed to do so by the owner, unless an emergency situation exists.

## 8.10 TRAFFIC CONTROL, FENCING, AND BARRICADING

Project locations may range from urban to rural in nature. Although specific needs will vary depending on the location, all projects require consideration of how the work will affect the adjacent community. Specific consideration must be given to pedestrian and vehicular traffic control.

During the design phase, traffic control or management of traffic plans may be developed and included in the bid documents. In these cases, the contractor implements the established traffic control plan in close coordination with the appropriate agencies. However, sometimes the contractor is fully responsible for traffic control and must develop its own plans in accordance with the needs of the local authorities, including the possible acquisition of permits to close travel lanes or sidewalks or to detour traffic. In either case, the plans typically follow the recommendations set forward in the *Manual of Uniform Traffic Control Devices* (MUTCD) published by the Federal Highway Administration (FHWA, 2012).

In addition to the traffic control around the work zone, it is important to protect pedestrians from the job site hazards. Protective measures typically include signage, barricades, fencing, and temporary pedestrian walkways.

Although the pilot tube and other guided boring methods are relatively quiet, shaft construction methods can be quite loud. Therefore, use of sound barriers to lower construction noise impacts to nearby sensitive receptors may be necessary. Typical receptors may include hospitals, schools, and residential structures.

## 8.11 QUALITY CONTROL

Quality control (QC) is an important part of the construction phase of the project. Controls that will help ensure the pipe is installed to acceptable line and grade include the following:

1. Require complete approved submittals prior to start of work.
2. Verify conformance with the approved submittals.
3. Use an experienced land surveyor to provide accurate construction layout.
4. Use equipment manufactured by a firm with a proven track record.
5. Require the pilot tube and other guided boring methods work to be performed by a qualified contractor as specified in the contract documents (refer to Chapter 5 for more information).

6. Verify line and grade of the initial pilot tube or guided bore pass before proceeding with subsequent phases. Reinstall if needed or otherwise adjust the installation of the next phase to improve the line and grade of an out-of-tolerance initial pass.

## 8.12 JACKING FORCES AND LUBRICANTS

The axial load exerted in each step of the pilot tube and guided boring process is the jacking force required to overcome the penetration resistance at the face of the pipe string plus the friction between the ground and the outside surface of the pipe being installed. A calculation of the anticipated jacking load should be performed prior to the start of the work to verify sufficient jacking capacity is available to complete the work. At no time should the jacking capacity of the tooling or the jacking pipe itself be exceeded.

There are two techniques to reduce the required jacking force. The first is to create an overcut, and the second is to introduce a lubricant to the outside surface of the pipe. In stable ground, an overcut reduces the friction between the pipe and the ground; a lubricant may or may not be needed. In less stable ground, the overcut will not remain open on its own, therefore a lubricant should be introduced into the overcut to stabilize the ground and reduce the friction.

Lubricants in pilot tube and other guided boring methods are used to reduce friction between the surrounding ground and the pilot tubes, temporary casings and/or jacking pipe. The selection of lubricants depends on the type of ground through which the pipe will be driven. The lubricant supplier should be consulted for guidance in determining the best lubricant to use for the project-specific geology, the proper method to mix the lubricant, and the most appropriate concentrations to achieve the desired results.

### 8.12.1 Use of Lubricants during Pilot Tube Installation

During the pilot tube installation, the steering head is usually larger than the outside diameter (OD) of the pilot tubes, thus as the steering head advances, it creates space (overcut) between the pilot tubes and the ground. In easily displaced ground, this may be sufficient to allow the pilot tubes to be installed for the full length of the drive without generating excessive installation loads.

If the ground is not easily displaced or if it is overconsolidated, or if sticky clays are present, lubrication may be required to reduce the jacking forces required to advance the pilot tube. The method for lubricant introduction varies by equipment manufacturer.

### 8.12.2 Use of Lubricants during Thrust Casing Installation

A reaming head adapter or other upsizing tool is attached at the rear of the pilot tube string to connect to the thrust casings. The reaming head is larger than the OD of the thrust casings, creating an overcut between the thrust casings and the ground. In stable ground and/or short drives, this overcut may be sufficient to allow the thrust casings to be installed for the full length of the drive.

Similar to the pilot tube advance, lubrication may be required to maintain the overcut and reduce the skin friction. In this case, lubrication is usually pumped through the pilot tubes from the receiving shaft, through integral ports in the upsizing tool, to the outside of the thrust casings. This will lubricate the leading edge of the casings as they are advanced.

### 8.12.3 Use of Lubricants during Final Jacking Pipe Installation

In the final installation step, when the jacking pipe OD matches the thrust casing OD, additional lubricant could be used to reduce jacking forces.

When the jacking pipe OD is larger than the thrust casing OD, an upsizing tool is used, such as a reaming head or a powered cutting head as described in Chapter 2. The reaming/cutting head is larger than the outside of the jacking pipe, creating an overcut between the pipe and the ground. In stable ground and/or short drives, this overcut may be sufficient to allow the final jacking pipe to be installed for the full length of the drive.

If the ground conditions are unstable, lubrication may be required to maintain the overcut and reduce the skin friction. In this case, lubricant is usually pumped through hoses running through the jacking pipe to ports on the reaming/cutting head, where it is injected into the overcut.

### 8.12.4 Use of Lubricants during Pullback of Final Pipe

After the pilot tube or thrust casing installation, the final pipe can be pulled back through the hole created. Additional lubricant can be used to reduce pulling forces.

## 8.13 SPOILS TRANSPORT AND DISPOSAL

The spoils generated through the boring process are conveyed to either the jacking shaft or the receiving shaft through the thrust casings for removal and disposal. The consistency of the spoils can range from dry soil to soupy mud (especially if the spoils have been lubricated). Usually, the spoils are removed from the shaft using a skip pan (dirt bucket) lifted to the surface with a crane or an excavator. An alternative to using a skip pan to remove the spoil from the shaft is a vacuum system.

It is important to ensure that the area around each shaft has sufficient space for spoils material stockpiling in addition to the safe operation of the support equipment. Often the excavated spoil is stockpiled in the staging area around the shaft for later hauling away and disposal. If the spoil is wet, containment may be required to avoid runoff onto adjacent properties. If contaminated soil is excavated, the owner should be notified, and the contractor should take the necessary measures for safe handling, storage, and disposal in conformance with the requirements of applicable regulatory bodies.

## 8.14 INSPECTION AND MONITORING

Implementation of the QC measures, outlined in Section 8.11, will ensure that the pilot tube and guided boring process is executed properly and will provide the highest likelihood of success.

Depending on the ground conditions, length of drive, diameter of jacked pipe, and proximity to adjacent structures, the engineer may require and specify implementation of a geotechnical monitoring program to ensure that the construction does not adversely affect adjacent structures or surrounding ground. These efforts may include

- Structural monitoring points (SMPs) installed to monitor for movement of nearby structures;
- Deformation monitoring points (DMPs) to monitor for ground movement (heave or settlement) along the pipe alignment;
- Utility monitoring points (UMPs) to monitor buried utilities for movement during construction;
- Piezometers or observation wells to monitor the groundwater table; and
- Inclinometers installed near the jacking and receiving shafts or near embankments to monitor for lateral ground movement.

Depending on the specific project conditions, the engineer may require a preconstruction survey to determine the existing conditions within the work zone. A preconstruction survey of nearby buildings and structures should document all existing cracks, cosmetic problems, and structural deficiencies. Further, elevations of SMPs, DMPs, UMPs, and piezometers should be surveyed and the inclinometers should be calibrated prior to shaft excavation to determine baseline readings of all instruments.

In addition to establishing the individual monitoring devices required and their locations, the monitoring plan should detail the surveying method to be used to take the readings as well as the frequency of monitoring. Along with the actual measurements taken, reports should include whether movement has occurred and if so, how much. The inspectors and the contractor can then take appropriate action to mitigate damage, if any. A monitoring program may also establish movement limits for the various monitoring points. Reaching these limits will trigger certain actions on the part of the construction team.

## 8.15 REPORTS AND RECORDS

Pilot tube and other guided boring equipment is typically not equipped with automatic data logging devices, so it is recommended that the boring activities be recorded manually on a paper form or computer tablet. The specifications should establish the frequency with which the selected operational functions are recorded in terms of elapsed minutes or number of times per pipe joint. The record should also include reference information, such as project name, operator's name, time, date, and pipe number. The operator should also note breakdowns and work stoppages, along with the reason for the breakdown or stoppage, in addition to observations of settlement, unusual events, and guidance system adjustments. An example is included in Fig. 8-7.

See Section 8.3 for a list of other information and data that could be included in a daily report as part of the construction submittals.

## 8.16 AS-BUILT DRAWINGS AND DOCUMENTATION

As-built drawings are generally required for any underground utility construction, including pilot tube and other guided boring methods projects. The contractor should mark the new pipe, manhole, and ancillary structure information on a copy of the contract plans. This process is known as redline drawings. The contractor should document any changes (vertical and horizontal) to the original layout of the underground utilities and structures that took place during the construction phase, such as rerouting a utility because of interference with the excavation of the shafts or changing the depth or alignment of the pipe to avoid difficult ground conditions. The redline drawings are usually submitted to the engineer for preparation of formal as-built documents for the owner of the facility.

**Pilot Tube Method Shift Report** Date: ________ Day: ____________

REPORT No. ______ Shift: Day ________ From ________ to ________

Project: ________________________ Eve ________ From ________ to ________

Contractor: ________________________ Night ________ From ________ to ________

Location: ________ Owner: ______________ Job No: ____ Pipe: ID: ________ OD: ________

FACE STATION: BEGIN __________ END __________ Length Jacked: ________ FEET

| Total Push Time | | List all activities performed while the excavation is (and is not) progressing | Pilots (PT) Cased Auger (CA) Product Pipe (PP) | Identification of Soil/remarks | Operator |
|---|---|---|---|---|---|
| Pipe No. | min | | | | |
| | | | | | |

OBSERVATIONS OF SETTLEMENT or HEAVE ______________________________

COMMENTS & INSTRUCTIONS FOR NEXT SHIFT ______________________________

REPORT PREPARED BY: ______________________ Signature: ______________________

*Fig. 8-7. Sample daily report form (courtesy of Northeast Remsco Construction, Inc.)*

## REFERENCES

FHWA (Federal Highway Administration). (2012). *Manual of uniform traffic control devices (MUTCD)*, Washington, DC.

OSHA (Occupational Safety and Health Administration). (1926). *Safety and health regulations for construction*, U.S. Dept. of Labor, Washington, DC.

# GLOSSARY

**Adapter ring**—A fabricated ring, usually made from steel, that is mounted on the first pipe of the pipe string. This ring is intended to transfer the jacking force from the boring equipment to the jacking pipe and to create a waterproof seal between the equipment and the spigot of the first joint.

**Advance rate**—Forward progress over a given period of time, which includes penetration rate, make-up time, down time, and idle time; typically measured in feet/shift or m/shift.

**Annular space**—The theoretical volume equal to annulus × length of tunnel.

**Annular thickness**—The radial distance between the excavated radius created by the gauge cutter and the outside radius of the jacking pipe. The radial distance is equal to the overcut (gauge cut) and the shield cut.

**Annulus**—The theoretical area between the gauge cut and the pipe outside diameter (OD) equal to $\pi \times (\text{Gauge Cut OD}^2 - \text{Pipe OD}^2)/4$. The area is the combined area created by the overcut and shield cut.

**Auger boring**—See **horizontal auger boring.**

**Auger casing**—See **thrust casing.**

**Auger MTBM**—A type of microtunnel boring machine, which uses auger flights to remove the spoil through a separate smaller casing placed through the jacking pipe. An auger MTBM maintains face support, is guided, and is remotely controlled.

**Backfill grouting**—Grout injected into the void space between the carrier pipe and casing after the drive is completed.

**Belly pans**—Shims or other steel plates added to the bottom of the boring equipment to adjust the location of the boring equipment within the excavation. Also known as *belly plates*.

**Berlin construction method**—A sewer collection system developed in Berlin, Germany, where sewer laterals are brought to a manhole and then dropped into the sewer interceptor. This method of connecting laterals to sewer is more compatible with trenchless installations.

**Boulder**—Per ASTM D653, a rock fragment, usually rounded by weathering or abrasion, with a dimension of 12 in. (305 mm) or more.

**Buried objects**—Items in the ground that can affect boring equipment advancement and line and grade control. Buried objects that prevent continued advancement along the design path including tolerances are obstructions.

**Caisson**—A watertight chamber used in shaft construction under water. The ground within the chamber is excavated, allowing the chamber to sink under its own weight into the ground.

**Carbon footprint**—The amount of carbon dioxide generated and caused by a construction project, including any traffic impacts.

**Carrier pipe**—Pipe for conveyance for water, gas, sewage, or other products and services. Some carrier pipe can be installed by direct jacking (i.e., one-pass installation).

**Cased auger**—See **thrust casing**.

**Cased bore**—A bore in which a pipe, usually a steel casing, is inserted simultaneously with the boring operation.

**Casing**—A direct-jacked pipe in a two pass system which supports a bore. The casing is not a carrier pipe.

**Cemented soils**—Soil in which the particles are held together by a naturally occurring or human-made chemical agent that changes the soils' physical and mechanical properties.

**Centerline**—a real or imaginary line through the center of something, especially one following an axis of symmetry. In a trenchless application, the centerline is through the vertical center of the pipe or boring equipment.

**Closed-face**—Term referring to one mode of operation of a traditional tunnel boring machine.

**Cobble**—Per ASTM D653, a rock fragment, usually rounded or semi-rounded, with dimensions between 3 and 12 in. (75 and 305 mm).

**Cohesionless soils (noncohesive soils)**—Earth materials containing less than 20% soil particles passing the #200 sieve. Any free-running soil, such as sand or gravel, whose strength depends on friction between particles.

**Cohesive soils**—Earth materials containing 20% or more soil particles passing the #200 sieve.

**Compression ring**—A ring fitted between the end-bearing area of the leading pipe bell and trailing pipe spigot to help distribute jacking forces more uniformly. The compression ring is attached to the trailing end of each pipe and is compressed between the pipe sections during jacking. The compression rings compensate for steering, pipe misalignment, and pipe dimensional tolerances. Also referred to as a *packer*.

**Contact grouting**—Grouting injected into the theoretical space between the jacking pipe and ground after the drive is completed.

**Contaminated plume**—The underground trace of an environmental pollutant.

**Contract documents**—Documents prepared by the owner and project engineer for bidding and for awarding a project; they can include bid forms, general conditions, special conditions, technical specifications, drawings, geotechnical data reports, and geotechnical baseline reports.

**Controls**—Part of the tunneling system that allows synchronized excavation, removal of spoils, and jacking of pipe to balance forward movement with excavation so that ground settlement and heave are managed.

**Crossing**—Pipeline installation in which the primary purpose is to provide one or more passages beneath a waterway, road, or other surface obstruction.

**Crown**—The highest point of the pipe or tunnel.

**Cut-and-cover**—See **open cut**.

**Cutability**—The ability to spall, rip, chip, and break rock into smaller pieces to allow the advance of the boring equipment. Key properties affecting cutability include the frequency of discontinuities, unconfined compressive strength, tensile strength, hardness, and abrasiveness of the rock.

**Cutter head**—Any rotating tool or system of tools on a common support that excavates at the face of a bore.

**Delay time**—Time associated with owner's issues.

**Deformation monitoring point (DMP)**—An instrument designed to measure ground movement along the pipe alignment during construction activities.

**Design specification**—Establishes specific requirements the contractor must use including the means and methods. A design specification creates an implied warranty that if the contractor uses the specified means and methods with industry-accepted good practices, the constructed product will meet the specification requirements.

**Desktop study**—A compilation and review of available site, geotechnical, and other information pertinent to a project. Typically performed at the outset of a project, prior to field explorations and detailed design.

**Dewatering**—The act of removing groundwater or lowering the groundwater elevation, using a system of wells and pumps.

**Differential cut**—The spiral, screw-like pattern created on the excavated surface as the gauge cutter is advanced while being rotated around the bored periphery of the cutter wheel.

**Differing site condition (DSC) clause**—A clause included in the contract documents to address how unanticipated ground conditions are handled.

**Dirt bucket**—See **skip pan**.

**Downtime**—Time lost when the boring equipment is unable to operate; generally associated with equipment or a system failure.

**Drilling fluid**—Water which may contain additives including bentonite, polymers, soda ash, surfactants, or other materials to enhance stability and excavatability. If the drilling fluid contains additives it is considered engineered drilling fluid, whether specified by the engineer or selected by the contractor. Some additives may increase the effectiveness of the drilling fluid and reduce adhesion of the spoils (cuttings). The drilling fluid is used in a closed-loop system for transporting spoils and counterbalancing earth and groundwater pressures during a trenchless installation. Typical trenchless methods that use drilling fluid include microtunneling and horizontal directional drilling.

**Drive**—Designation of the pipe installed from a jacking shaft to a receiving shaft.

**Drive shaft**—See **jacking shaft**.

**Earth piercing**—The use of a tool which comprises a percussive hammer within a suitable casing, generally of torpedo shape. The hammer may be pneumatic or hydraulic. The term is usually associated with non-steered devices.

**Earth plug**—Intentional buildup of excavated material at the leading edge of the boring equipment, auger, or casing; intended to improve face stability.

**Effective stresses**—In a saturated mass of soil or rock, the total stress $P$ minus pore water pressure, $U_w$.

**Emergency recovery shaft**—See **rescue shaft**.

**Entrance seal**—See **launch seal**.

**Entry ring**—See **launch seal**.

**Entry shaft**—See **jacking shaft**.

**EPB machine**—Earth pressure balance tunneling machine in which mechanical pressure is applied to the material at the face and controlled to provide the correct counter-balance to earth and groundwater pressures to prevent heave or subsidence. The term is not applicable to **microtunneling** systems for which the primary counterbalance of earth and groundwater pressures is supplied by pressurized slurry.

**Exit seal**—Same as **launch seal** except for the retrieval of the boring equipment at the receiving shaft. Used in high groundwater and unstable soils to prevent loss of ground.

**Exit shaft**—See **receiving shaft**.

**Face**—The location where excavation is taking place.

**Face pressure**—Earth and groundwater pressures applied against the cross sectional area of the bore or tunnel face.

**Factor of safety**—A ratio between the structural capacity of a system divided by expected loads or actual loads.

**Filter cake**—A thin layer of clay or polymer from the drilling fluid and perimeter of the formation being excavated or displaced. The filter cake is formed through filtrate loss.

**Fog**—Moisture or dust that exists in the pilot tube that impedes the light-source guidance.

**Frac-out**—See **inadvertent returns**.

**Gauge cut**—The outermost projection of the excavation, measured as a diameter. The gauge cut is subject to cutter wear and the differential cut.

**Gauge cutter**—The cutter that creates the gauge cut and is subject to wear.

**Gel**—Per ASTM D653, the condition where a liquid begins to exhibit measurable shear strength.

**Gel time**—The measured time interval between the slurry mix in a fluid state and the formation of a gel.

**Gravel**—Per ASTM D653, rounded or semi-rounded particles of rock that will pass a 3-in. (76.2 mm) and will be retained on a No. 4 (4.75 mm) US standard sieve.

**Ground cover**—Distance between the ground surface and the crown of the pipe.

**Groundwater**—All subsurface water, as distinct from surface water.

**Grouting**—The process of filling voids or modifying/improving ground conditions. Grouting materials may be cementitious, chemical, or other mixtures.

**Guidance system**—System that locates the actual position of the boring equipment relative to the laser, light source, or other device. The laser, light source, or other device should be referenced to the design line and grade.

**Guided boring**—A general term used to describe various trenchless methods of installing a pipe to a desired line and grade using a guidance system. The boring process does not provide continuous support at the face of the excavation to balance groundwater and earth pressures.

**Heave**—Measurable upward movement of the ground or structure as the result of the excavation process.

**Horizontal auger boring**—A technique for forming a bore from a jacking or drive shaft to a receiving shaft by means of a rotating auger with cutting tools. The casings are jacked forward sequentially in a cyclic process while turning the auger. Spoil is removed back to the drive shaft by helically wound auger flights rotating in the steel casing.

The equipment may have limited guidance and steering capability.

**Horizontal directional drilling**—A surface-launched trenchless technology for the installation of pipes, conduits, and cables. Horizontal directional drilling, also called HDD, creates a pilot bore along the design pathway, and reams the pilot bore in one or more passes to a diameter suitable for the insertion of the carrier pipe. The carrier pipe is pulled into the prepared bore in the final step of the process. HDD is typically used in nongravity flow applications.

**Idle time**—Time lost when the boring equipment is not operated, not including make-up time or down time.

**Inadvertent returns**—The loss of drilling fluid, including slurry and lubrication, from the slurry or lubrication system. A special form of inadvertent return, where the fluid exceeds the strength and confining pressure of the ground and reaches the surface or waterway, is called a hydro-fracture or **frac-out**.

**Inclinometer**—(1) A geotechnical instrument used to measure horizontal deflection transverse to an installed vertical pipe casing by passing a probe through a casing (via two sets of grooves in the casing). (2) An instrument used within the boring equipment to measure pitch and roll.

**In-line microtunneling**—A method of replacing an existing pipe by filling the pipe with flowable fill to prevent fluid loss and then excavating the entire pipe and surrounding ground with an MTBM for a new installation.

**Interjack pipes**—Pipes specially designed for use with an **intermediate jacking station**.

**Intermediate jacking station**—A fabricated steel cylinder fitted with hydraulic jacks, which is incorporated into a pipe string between two specially fabricated pipe segments. Its function is to provide additional thrust to overcome skin friction and to distribute the jacking forces over the pipe string on long drives.

**Jacking forces**—The total force required to overcome the face pressure component and frictional resistance component along the pipe to allow the forward movement of the boring equipment and trailing pipe string.

**Jacking frame**—A structural component that houses the hydraulic cylinders used to propel the boring equipment and pipeline into the ground. The jacking frame cradles the boring equipment and jacking pipes and serves to distribute the thrust load to the pipeline and the reaction load to the shaft wall or thrust block.

**Jacking loads**—See **jacking forces**.

**Jacking pipes**—Pipes designed to be installed using **pipe jacking** techniques.

**Jacking shaft**—Excavation from which **trenchless technology** equipment is launched for the installation of a pipeline. The jacking shaft may incorporate a thrust wall to spread reaction loads to the ground and an **entry ring** to control inflows of groundwater and earth at the portal.

**Jacking shield**—A fabricated steel cylinder from within which the excavation is carried out either by hand or machine. An articulated steering section and steering jacks are incorporated within the shield to allow it to be adjusted to control line and grade. Term is associated with traditional tunneling.

**Laser**—A device commonly incorporated in the guidance system used to track alignment and grade during the tunneling or jacking operation.

**Launch seal**—A mechanical seal, usually a rubber flange that is mounted to the wall of the jacking shaft. The flange seal is distended by the boring equipment as it passes through. The seal is intended to prevent groundwater, soils, slurry, and lubrication inflow into the shaft during tunneling operations.

**Launch shaft**—See **jacking shaft**.

**Lubrication**—The act of injecting a fluid, normally bentonite and/or polymers, to reduce the skin friction and **jacking forces** on the jacking pipe during installation. The fluid fills the **annular space**.

**Make-up time**—The time spent to add a pipe segment and make connections.

**Marsh funnel viscosity**—A measure of the flow rate of a fluid through a calibrated funnel. Marsh funnel viscosity is typically used to measure drilling fluid and slurry viscosity, in seconds per quart of fluid. The Marsh funnel viscosity of water is 26 seconds per quart.

**Maximum allowable jacking force**—The largest jacking force that can be applied to the jacked pipe, allowing for an appropriate factor of safety.

**Maximum anticipated jacking force**—The largest anticipated jacking force required to advance jacking pipe and boring equipment along a drive.

**Microtunneling**—A trenchless method to install pipe into the ground that meets four specific features during construction: (1) Remote controlled: The microtunneling boring machine (MTBM) is operated from a control panel, normally located on the surface. The system simultaneously installs pipe as the ground is excavated and removed. Personnel entry is not required for routine operation. (2) Guided: The guidance system usually references a laser beam projected onto a target in the MTBM, capable of installing gravity sewers or other types of pipelines to the required tolerance for line and grade. (3) Pipe jacked: The process of constructing a pipeline by consecutively pushing pipes and MTBM through the ground using a jacking system for thrust. (4) Continuously supported: Continuous pressure is provided to the face of the excavation to balance groundwater and earth pressures.

**Mixed face**—An interface within the excavated tunnel zone between two geological units with significant contrast in engineering properties, e.g., rock overlain by soft ground or very soft, low-strength soil overlain or underlain by a very stiff, high-strength soil.

**Mixed ground**—The excavated tunnel zone without a defined interface that contains soil and coarse fractions of **gravels**, **cobbles**, and **boulders**, i.e., silt with boulders. This term is not currently defined by ASTM or USCS.

**Mixed reach**—A significant change in ground type, characteristics, or behavior along the excavated tunnel between the jacking and receiving shafts.

**Muck**—Spoil or removal of same.

**Muck bucket**—See **skip pan**.

**N-value**—The number of blows per foot required to drive a standard penetration test (SPT) soil sampler into the ground during geotechnical exploration (ASTM D1586). The N-value is often used to estimate in situ consistency of clays and silts and relative density of sands.

**Noncohesive soils**—See **cohesionless soils.**

**Observation well**—See **piezometer.**

**Obstruction**—Any buried object that lies completely or partially within the cross-section of the boring equipment that impedes continued forward progress along the design path and within allowable tolerances.

**One-pass**—The carrier pipe is directly jacked into the ground.

**Open cut**—A method by which access is gained to the required level underground for the installation, maintenance or inspection of a pipe, conduit, or cable. The excavation is then back-filled and the surface reinstated.

**Open face**—Term which refers to a mode of operation of a traditional tunnel boring machine.

**Overcut**—The theoretical difference of the radial measurement of the gauge cut and the boring equipment shield, equal to (Gauge Cut OD – shield OD)/2. Actual overcut is reduced as the gauge cutter is worn and because of the differential cut.

**Packer**—See **compression ring**.

**Penetration rate**—Instantaneous excavation distance per time while the boring equipment is operating, typically measured in inches/min or mm/min.

**Perched water**—An accumulation of groundwater that is above the regional groundwater table in an unsaturated zone.

**Performance specification**—Establishes the performance criteria the constructed product must meet. A performance specification leaves the selection of the means and methods to the contractor.

**Piercing tools**—See **earth piercing.**

**Piezometer**—A specialized monitoring well that provides discrete screens and underground seals to provide a measure of the groundwater pressure at specific intervals below ground. The groundwater pressure may be measured using open stand pipes or with vibrating wire transducers.

**Pilot tube**—A small-diameter, hollow, stiff-walled pipe string (tube). Typically, a light source is mounted at the leading edge and pushed through displaceable ground. Insertion of the tube is the first step in the installation process.

**Pilot tube method**—A multistage method of accurately installing a pipe to line and grade by use of a guided pilot tube followed by upsizing to install the pipe.

**Pipe brake**—See **pipe clamp**.

**Pipe clamp**—A mechanical device designed to prevent the boring equipment and pipe string from moving back into the shaft. The device is also called a pipe brake.

**Pipe eating**—See **in-line microtunneling**.

**Pipe jacking**—A system of directly installing pipes behind a shield machine by hydraulic jacking from a drive shaft such that the pipes form a continuous string in the ground.

**Pipe lubricant**—See **lubrication**.

**Pipe ramming**—A trenchless installation method whereby a percussive hammer is attached, via an adapter, to an open-ended casing, which is driven through the ground. The spoil within the casing is removed after the drive is completed or periodically during the drive, to create an open casing.

**Pipe string**—The succession of joined individual pipes being used to advance and support the excavation.

**Pit**—A vertical hole shallow shaft usually less than 5 ft used for launching or receiving the trenchless equipment and commonly may not require wall support system.

**Pitch**—The upward or downward angle of the boring equipment measured from a theoretical horizontal plane passing through the longitudinal axis of the boring equipment.

**Porewater pressure**—The pressure of the groundwater held within a soil or rock, in gaps between particles (pores).

**Port**—Opening in the barrel of the pipe used to introduce fluids (lubricant and/or grout) into the annular space between the pipe and excavated ground.

**Potholing**—Small, carefully dug, nonmechanized excavation to locate a utility or other subsurface feature.

**Product pipe**—See **carrier pipe**.

**Pullback**—A step in the installation process where the carrier pipe is pulled into the created hole.

**Push ring adapter**—Mechanical component mounted on the thrust ring to prevent the thrust ring from coming in contact with the pipe collar and causing damage to the collar.

**Receiving shaft**—Excavation into which the boring equipment is driven and recovered.

**Reception shaft**—See **receiving shaft**.

**Rescue shaft**—An unplanned additional shaft required to remove obstacles/obstructions and/or retrieve or repair the MTBM. The rescue shaft may need to function as a jacking shaft to complete the drive.

**Rider**—A shallow collection line which runs along the trenchless trunk sewer to a drop manhole.

**Rock**—Per ASTM D653, any naturally formed aggregate of mineral matter occurring in large masses or fragments (ISRM).

**Roll**—The angle of rotation about the theoretical longitudinal centerline of the boring equipment measured from the 12 o'clock position.

**Settlement**—Measurable downward movement of the ground, overlying utility, or other structure as the result of the excavation process or dewatering.

**Settlement point** or **marker**—An instrument installed at ground surface or driven into the ground to monitor vertical ground deformations before, during, and after construction.

**Shaft**—A deep pit (vertical excavation) usually more than 5 ft for launching and receiving boring units.

**Shield cut**—The theoretical difference of the radial measurement of the shield and the jacking pipe, equal to (shield OD: Pipe OD)/2.

**Skinning**—Also referred to as upsizing or shimming, the act of fitting a steel cylinder over the boring equipment shield body to increase the diameter of an existing boring tool with use of an appropriately upsized cutter wheel.

**Skip pan**—A bucket used in the shaft to collect and hoist spoils from the tunnel for removal. Also known as a dirt bucket or a muck bucket.

**Slurry**—A mixture of drilling fluid and spoils.

**Spacers**—Mechanical structure used to transfer the **jacking force** from the jacking **thrust ring** to the pipe and used to accommodate lengths of pipe that are longer than the stroke length of the jacks.

**Specials**—The pipe sections immediately ahead of and behind the intermediate jacking station (IJS) that have specially manufactured ends to physically accommodate the IJS.

**Spoil**—Earth, rock, and other materials excavated during the installation process. Also referred to as cuttings.

**Structural monitoring point (SMP)**—An instrument designed to measure ground movement of nearby structures during construction activities.

**Sump pump**—A device placed at the base of a shaft or in the boring equipment to collect and remove fluids incidental to the construction process and shaft leakage, and to prevent the excavation equipment from flooding.

**Thrust block**—An engineered structure located between the **jacking frame** and the shaft wall which distributes the **jacking force** developed by the hydraulic jacking frame over a large surface area to the ground behind the shaft back wall.

**Thrust casing**—A fabricated casing with an auger inside to enlarge the hole created by the **pilot tube**, typically associated as the second step in the installation process and fully retrievable.

**Thrust ring**—A fabricated ring that is mounted on the face of the **jacking frame**. It is intended to transfer the jacking force from the jacking frame to the thrust-bearing area of the pipe section being jacked.

**Tooling**—Ground-engaging elements of the cutter wheel designed to excavate and penetrate the ground.

**Top hat**—A steel can with a face plate bolted to a receiving shaft wall to receive the arriving boring equipment.

**Trenching**—See **open cut**.

**Trenchless technology**—A family of construction techniques for installing or rehabilitating underground infrastructure with minimal disruption to surface traffic, businesses, and residents. Also includes technologies for inspection, leak location and detection, with minimal disruption and excavation from the ground surface.

**Tunnel horizon**—The vertical band within the ground where the tunnel is excavated; includes one tunnel diameter above the crown and one tunnel diameter below the invert.

**Two-pass**—The carrier pipe is installed within a **casing**.

**Utility monitoring point (UMP)**—An instrument designed to measure movement of an individual utility when exposed to construction activities.

**Utility tunneling**—A construction method of excavating an opening beneath the ground without continuous disturbance of the ground surface. The excavation is of sufficient diameter to allow personnel access, excavation, transport of spoils, and erection of a ground support system.

**Unanticipated returns**—See **inadvertent returns**.

**Upsizing**—Process of creating a larger-diameter boring equipment than the original factory configuration. Also, any method which increases the cross sectional area of an existing pipeline, replacing it with a larger cross sectional area pipe.

**Upsizing tool**—A fabricated cutting tool such as reaming head or a powered cutting head used to enlarge the hole created by the **pilot tube** or **thrust casing**. The upsizing tool allows the jacking pipe to be installed directly behind the tool as the pilot tube or thrust casing is displaced.

**Water jetting**—Internal cleansing mechanism of the **cutter wheel** in which high-pressure water is sprayed from nozzles to help remove **cohesive soils**.

**Yaw**—The angle of the boring equipment to the left or right of a theoretical vertical plane passing through the longitudinal boring equipment axis.

**Zone of influence**—Volume of ground that could possibly be impacted by settlement or heave from either the tunnel mining, shaft excavation, pile driving, or other construction activities.

# INDEX

**Page numbers followed by *e*, *f*, or *t* indicate equations, figures or tables.**